Data Acquisition Systems

Data Acquisition Systems

Arnold H. VanDoren

RESTON PUBLISHING COMPANY, INC.
A Prentice-Hall Company
Reston, Virginia

Library of Congress Cataloging in Publication Data

VanDoren, Arnold H.
Data acquisition systems.

1. Automatic data collection systems. I. Title.
TS158.6.V36 621.3819'5 82-328
ISBN 0-8359-1216-7 AACR2

Interior design and editing/production supervision
by Ginger Sasser

10 9 8 7 6 5 4 3 2 1

Printed in the United States of America

Table of Contents

Preface

During the past 15 years, the world has witnessed an evolution in the measurement of analog information for computer processing. This has been brought about by significant advances in semiconductor technology and by the availability and proliferation of inexpensive digital computer hardware. A whole new industry has developed and flourished. In spite of this dynamic growth, there is a substantial lack of textbook material which puts all the various aspects of data acquisition together in one volume. Selection of a textbook for the course I have taught for the past several years has been most frustrating. While there are many good works covering the individual aspects of data acquisition, no one book, or even two books, covers the broad spectrum of the subject.

This book has been written to help solve that problem. It is presented as an introduction to the field of data acquisition and as a useful tool for engineers who are entering the field. Every attempt has been made to present the many complex concepts in a way that is easy to understand and which makes them readily usable. The questions at the ends of the chapters have been selected to support the text and not to confuse the reader.

The book is organized so that the reader will work his way from the analog input of the data acquisition system, through the various signal conditioning and conversion functions to the digital computer. As a preface to the analog functions, the reader will investigate the many problems associated with sampling, quantization and subsequent recovery of the analog signals. He is also acquainted with the specifications and parameters which are characteristic of the data measurement field.

I would like to acknowledge the help of my secretary, Joyce McNeil, who spent many hours typing this manuscript, and for the patience and understanding of my wife, Barbara, for the many inconveniences the work on this book created.

Arnold H. VanDoren

1

The Data Acquisition System

1-0 WHAT IT IS

Today's world functions almost totally on decisions that are based on the gathering of data and its subsequent analysis. A child puts a foot into the water to determine the temperature before diving in. Life insurance firms gather huge volumes of actuarial data to determine risks and to establish premiums. There are nearly as many systems for acquiring data as there are types of data. This text, however, confines itself to a particular class of *data acquisition system*, defined as:

> an electronic instrument, or group of interconnected electronic hardware items, dedicated to the measurement and quantization of analog signals for digital analysis or processing.

In this context, the data acquisition system (DAS) is the analog interface to the digital world. A graphic representation of where the data acquisition system "fits in" is found in Figure 1-1. Once the parameter to be measured is translated into the analog-electrical domain, the DAS performs the translation to the digital-electrical domain. In some cases the DAS simply records, or stores, the digital data, while more

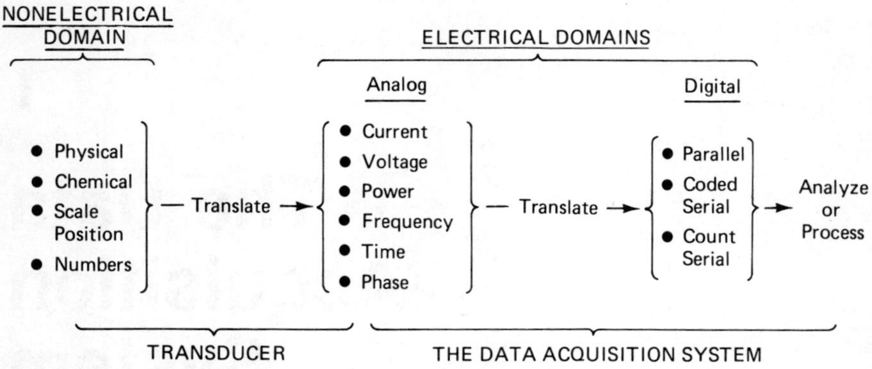

Figure 1-1. Where the Data Acquisition System Fits

sophisticated systems may be capable of analysis or further processing. For instance, a DAS may be as simple as a digital voltmeter (DVM), which displays its output as a decimal readout, or it may be complex enough to contain a large-scale computer as part of its hardware.

Now that we understand the function of this type of system, let's examine the components of a basic data acquisition system, as shown in the block diagram in Figure 1-2. Each of the blocks in this figure represents a particular data acquisition function, and each is defined in general terms as follows:

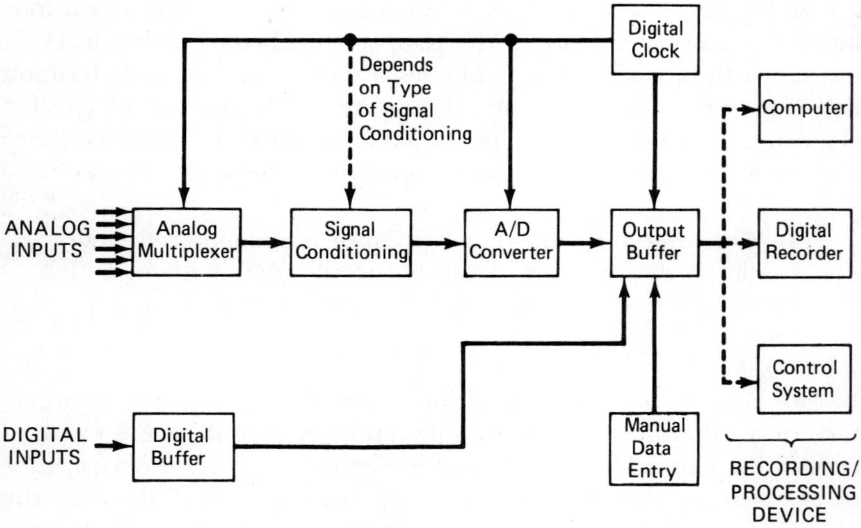

Figure 1-2. Basic Data Acquisition System—Block Diagram

Analog Multiplexer

The analog multiplexer permits a number of signal sources to be automatically measured by the same data acquisition hardware. It consists of a series of switches whose inputs are tied to the various analog signals and whose outputs are tied to a common measuring point. Each input is individually connected to the measuring point in a predetermined sequence. The number of channels in a multiplexer may vary from two to several hundred.

Signal Conditioning

Very often the signals presented to the inputs of the data acquisition system are not in a form appropriate for conversion, and so they must be preconditioned. The required signal conditioning could consist of linear amplification, logarithmic amplification, filtering, peak detection, or sample-and-hold. Often more than one of these functions is required. For instance, it is not uncommon to combine amplification with filtering or to find a low-level amplifier before a sample-and-hold.

Analog-to-Digital (A/D) Converter

The analog-to-digital converter actually translates the analog signal into an encoded digital format. Of the numerous ways to perform this function, only about a half-dozen techniques have found wide acceptance. Most notable are the dual-slope integrating and the successive approximation converters. A/D converters are often referred to by the number of output digits they produce. In a binary system, the range is from 4 to 16 bits, while in a binary-coded decimal system, 3 to 4 digits are normal.

Digital Clock

The digital clock provides the master timing for the data acquisition system. It may be as simple as a multiphased crystal controlled oscillator, or it may provide the user with a wide selection of multiplexer rates and modes of operation. Some systems also contain both time-of-day and day-of-year clocks.

Manual Data Entry

Many data acquisition systems provide users with a way to tag the data they are accumulating, through some sort of manual data entry. They may wish to note such things as the type of data, special conditions surrounding the measurements, and/or the run number. Such information is generally provided in a "header" that precedes the actual measurement data.

Digital Buffer

The ability to record discrete events is often a requirement of a data acquisition system. Since these events are usually accompanied by the opening or closing of a switch, they represent a digital input. Out-of-tolerance conditions or some other situations that might invalidate the data collection are most often designated as "discrete events."

Output Buffer

The output buffer acts as the data collector for the DAS. In an ordered sequence, it gathers up such data as the multiplexer channel number, the signal conditioner gain, A/D converter data, manual data, clock information, and discrete events. The buffer combines the data with the proper format for entry into the recording or processing system. It also provides the proper buffering and control to interface with the recording or processing device.

If the processing device were a minicomputer, the output buffer might be called a "peripheral controller."

Recording/Processing Device

A number of different equipment types can fill the role of the recording or processing device. Some of the equipment types commonly used are paper tape punches, teleprinters (TTY), magnetic tape units, line printers, cathode-ray tube displays, floppy disks, general-purpose digital computers, and special-purpose digital processors.

In recent years, all these components of the data acquisition system have become more automated; a typical block diagram can be seen in Figure 1-3. All the major functions of Figure 1-2 can be found with the exception of manual entry, which is accomplished directly by the computer through the keyboard display. Clock information is supplied by the computer, and timing is generated in the input/output (I/O) controller. This system configuration has two distinct advantages: first, on-line programmable processing is possible; and second, a host of storage media is available. The rapid expansion of the data acquisition field is directly attributable to the advent of low-cost computer hardware, which makes systems like that of Figure 1-3 a reality. Yet we should not overlook the dotted-line box in the lower right-hand corner of that figure: Software can easily become a more costly burden than the hardware.

The latest trend in test instrumentation is to embed a micro-

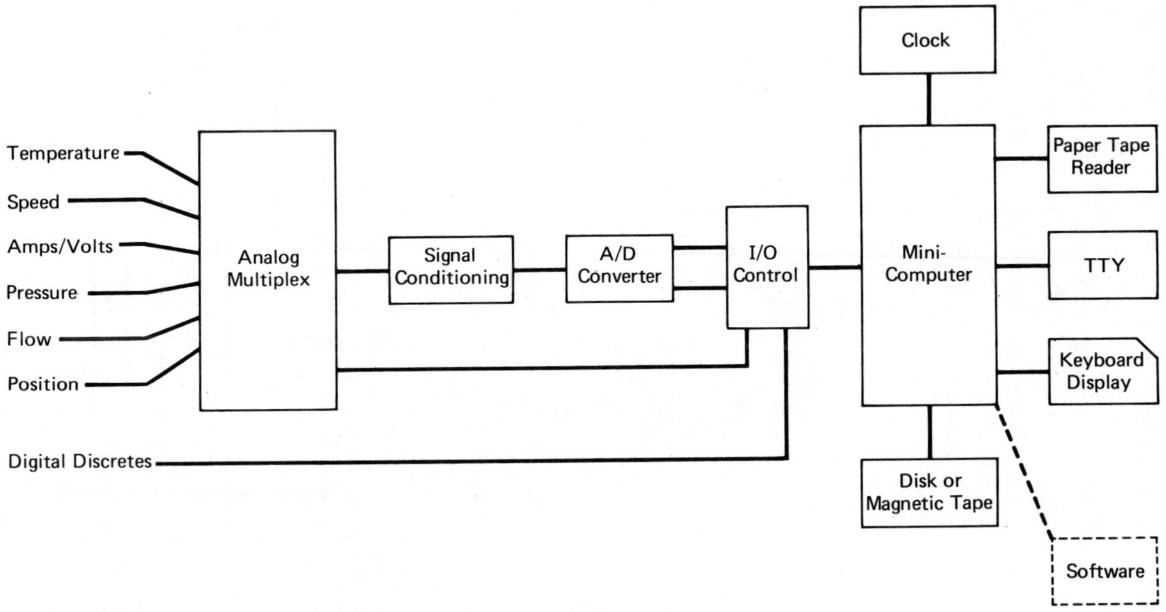

Figure 1-3. More Automated Version of DAS

processor in the heart of the equipment, as shown in Figure 1-4. This places tremendous programming power and flexibility in the hands of the user.

Other data acquisition configurations are also worthy of note. Figure 1-5 shows a system that utilizes a variety of analog-to-digital converter types. Such a system would apply to a situation in which the test data comes from a variety of different sources. The output buffer must also perform the function of digital multiplexing.

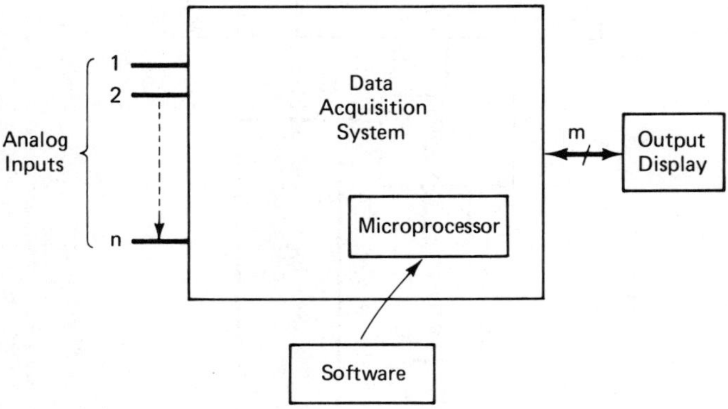

Figure 1-4. Microprocessor-Controlled DAS

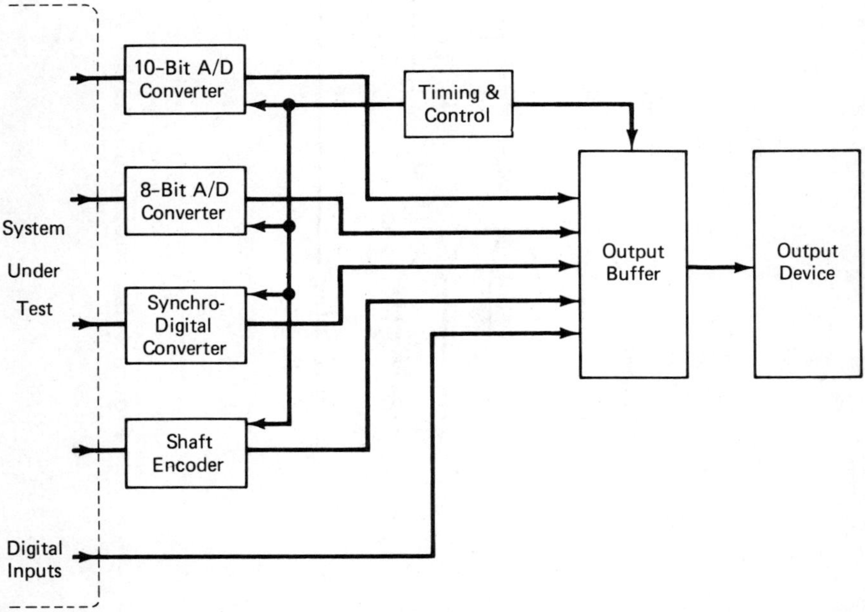

Figure 1-5. DAS with Multiple Encoders

A totally different configuration is found in Figure 1-6. Here each channel has its own signal conditioning and A/D conversion. This system concept is usually employed when the analog signals come from a number of very diverse sources and/or when high calibration accuracy is required. Each channel may be separately aligned. A few years ago, this approach would have been an extremely expensive proposition. With the decrease in the cost of DAS hardware, however, it is rapidly becoming more and more viable.

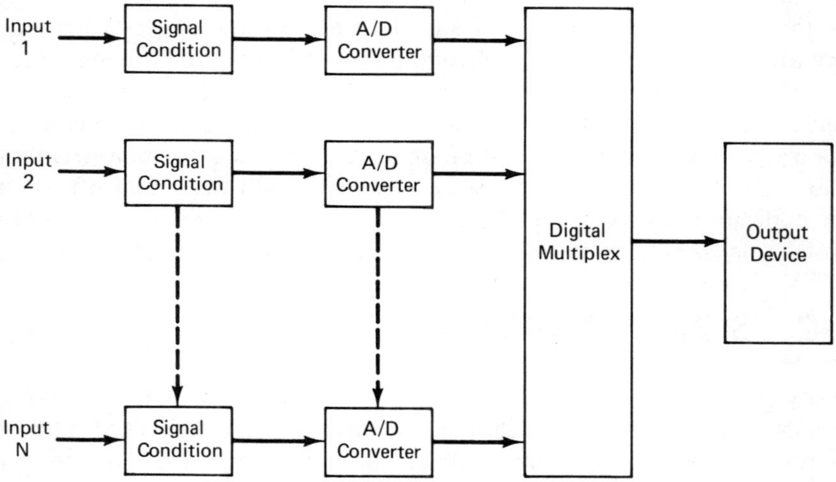

Figure 1-6. DAS with Conditioning and Conversion for Each Input

Having carefully examined the data acquisition system itself, let's turn our attention to some of its potential uses. Although any hardware system that has both an analog and a digital nature will have a DAS embedded in it somewhere, we will restrict our discussion to four broad areas. They are:

1. data logging,
2. signal analysis,
3. automated factory testing, and
4. process control.

1-1 DATA LOGGING

The *data logger* is a data acquisition system that measures the analog inputs, translates the results into the digital domain, and stores the data for future processing or analysis. Figure 1-2 depicts the typical data logger system block diagram. From the earliest days of recorded data, scientists have set up experiments with appropriate analog measuring devices at critical points in the operation. They have tabulated their data either by visual observation at set intervals or by recording it on another analog device, such as a pen recorder. In either case, they had to spend many hours analyzing the results. The data logger provides an automated method of making the measurements and recording the data. It can perform these functions at precise intervals and with a degree of accuracy beyond any person's physical capability. Because the data is stored in digital format, the data analysis is easily performed by an off-line computer. Since many data loggers are highly portable, data-taking can be automated even in very remote locations.

Most logging systems require very little attention. After the experiment is set up and the data logger started, the operator may leave the measurement site. The data logger will carry out the measurement plan precisely as the operator set it up, and it will shut itself off when the experiment is complete. As long as we continue to explore the world around us, there seems to be no limitation for this type of hardware.

1-2 SIGNAL ANALYSIS

Many processes in our world today are characterized by time-varying signals from which much useful data may be extracted. Examples of such processes are signature analysis, radar tracking and warning, air traffic control, electronic countermeasures, seismic exploration, and patient monitoring. The most common way to process such information is to translate it into an appropriate Fourier series or transform for analysis in the frequency domain. Low-cost computer and processing hardware, coupled with computationally efficient fast Fourier transforms (FFT), has made this approach feasible. Since the FFT algorithms work with discrete signal samples represented in digital form, the front end of a system designed for signal analysis must be a typical data acquisition system. The growth of this area is bounded only by a few things: the cost of computer storage, along with the operating speed of computer and data acquisition hardware.

1-3 AUTOMATED FACTORY TEST EQUIPMENT

Large-volume manufacturing has provided another lucrative area for the data acquisition system. With an increase in production quantities, testing time—as well as the burden of training a large number of qualified test personnel—must be reduced. A desirable test system would require only inserting the product into a fixture, setting at most two or three simple switches, and depressing a test button. The test set would automatically exercise the product, quantize the results, compare the results against the desired answer, and activate a go or no-go signal as required. Hard copy should be available if necessary.

The heart of such a test system is, of course, our old friend the data acquisition system. Automated test systems often include a small, general-purpose digital computer. The computer provides inputs to exercise the product, stores or records the test information, and examines and analyzes the results. In some factories, the test system is so sophisticated that several test positions containing small computers are tied together by a larger central processor, which stores and analyzes data from and maintains the status of all test positions.

1-4 PROCESS CONTROL

Several years ago, process control systems were essentially human-controlled. Operators established the process parameters by means of analog set-point controllers, and they monitored results by means of visual recorders and annunciators. Simple, fixed control algorithms were generally used, and only a limited number of points could be monitored and controlled.

Today, the data acquisition system has become a way of life in most types of industrial process control. The first data acquisition system applications to process control came when a computer was added to monitor the process. Not only could the computer now monitor several hundred inputs, but it could also display the readings, make calculations, and provide correction data to the operator. This concept still required operators to take action, based on the computer output, but it gave users a comforting feeling that their process was not subject to the whim of a piece of "unreliable" computer hardware.

As computer hardware has become more efficient and reliable, the process control system has evolved into that pictured in Figure 1-7. The DAS and computer are clearly parts of the control loop as it measures the key parameters, makes reference comparisons, computes the control algorithm, and provides feedback correction to the process. The operator is now the monitor—interrogating the system through the operator console, checking the computer-displayed indicators, and answering alarms.

1-5 WHAT COMES NEXT?

The remainder of this text is designed to acquaint you with the principles of data acquisition systems. Chapters 2 and 3 cover the basics of sampling and quantization theory. Chapter 4 describes the basic DAS

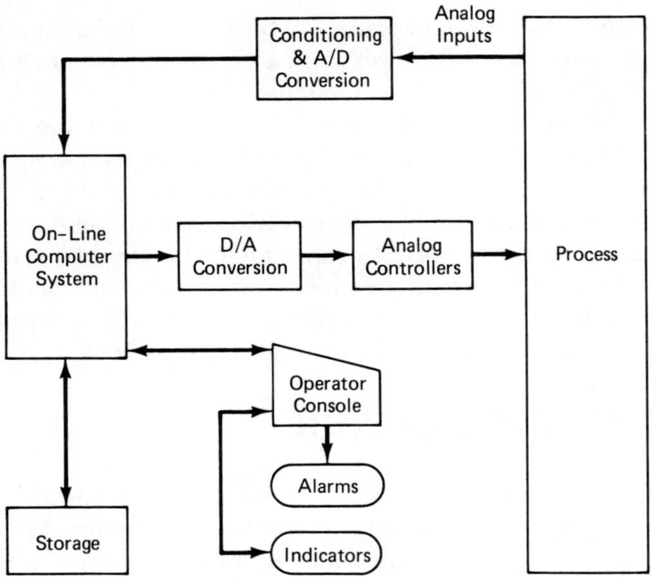

Figure 1-7. Computer-Automated Process Control

parameters and specifications. Each of the basic functions that entail analog processing is discussed in detail in Chapters 5 through 8. Because nearly all the error or inaccuracy in a DAS arises from areas in which the information either is in analog form or is being translated from or to the analog domain, Chapter 9 provides an insight into error analysis. With the analog side of the system completed, Chapters 10, 11, and 12 consider the digital portions of the system.

REFERENCES

1. P. H. Garrett, *Analog Systems for Microprocessors and Minicomputers*. Reston, Va.: Reston Publishing Company, Inc., 1978.
2. H. N. Norton, *Handbook of Transducers for Electronic Measuring Systems*. Englewood Cliffs, N.J.: Prentice-Hall, Inc., 1969.
3. M. Schwartz and L. Shaw, *Signal Processing: Discrete Spectral Analysis, Detection, and Estimation*. New York: McGraw-Hill Book Company, 1975.
4. W. Stanley, *Digital Signal Processing*. Reston, Va.: Reston Publishing Company, Inc., 1975.

2
Basic Sampling Concepts

2-0 INTRODUCTION

Fundamental to the operation of a data acquisition system is the concept of *sampling*, which may be defined as:

> the act of measuring a continuous function at discrete time intervals.

Signals that represent some physical or analog parameter are converted to a series of discrete values. Unfortunately, many instrumentation engineers are content when they think they understand only the rudiments of the sampling process. As it turns out, since the data acquisition system is a sampled-data system, the overall system transfer function must be examined to achieve a proper *solution*.

A typical data acquisition/sampled-data system is shown in Figure 2-1. As the diagram indicates, four major processes are involved:

1. sampling,
2. quantization,
3. digital processing, and
4. recovery.

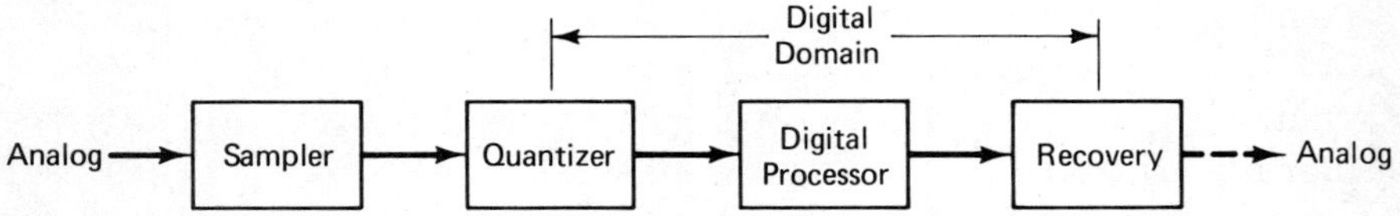

Figure 2-1. Typical Data Acquisition/Sampled-Data System

A properly designed system generally requires balancing the implementation of one of these processes with the characteristics of the others. For instance, the sample rate may be a compromise between the complexity of the sampler and the sophistication of the recovery filter. If the required accuracy and/or system dynamic range is high (a large number of quantizer levels), both the sample rate and the recovery process may be affected. The required number of quantization levels may also be dictated by the digital processing algorithm.

This chapter considers the fundamentals of sampling, while Chapter 3 is devoted to quantization and signal recovery. To round out the discussion, Chapter 3 ends with a practical example and a summary of design considerations.

Besides the required number of bits (or quantization levels), the only major problems in a data acquisition system involve first the proper conversion of analog signals to digital form, and then the subsequent reconversion of the original signal, or its derivative, back to analog. (Consideration of the various digital processing algorithms, however, is beyond the scope of this text.)

2-1 KINDS OF SIGNALS

Before applying the concepts of sampling, quantization, and recovery, you must have a reasonable understanding of the limitations of the signal to be processed. Is the signal dc, quasi-dc, or dynamic? Is it deterministic or random? What are the dynamic range, the frequency range, and the noise characteristics? What is the shape of the frequency spectrum? These questions, along with the information to be acquired, are fundamental to the design of a data acquisition system.

We will now examine some of the signal types that might be encountered.

2-1.1 Static or dc Signals

A *static* or *dc signal* is one that is stable or not changing. The first question you might ask is, "Why would you want to continuously measure an unchanging signal when one measurement would be sufficient?" In many systems the dc power source(s) may require monitoring. Generally, the measurement system records the same constant value over and over. Yet, should the voltage drop out of tolerance, an alarm is triggered so that corrective action may be taken. Often, this situation is compounded when an undesirable noise signal is superimposed on the dc signal. The results depend greatly on the type of measurement hardware and on the occurrence of the noise, relative to the measurement point. Should the measurement be made coincident with a noise peak, the reading could deviate drastically from the expected result.

One solution to this problem is *signal averaging*. In this approach, a large number of measurements of the combined dc/noise signal are made, and their sum is divided by the number of such measurements. Thus the signal-to-noise ratio is improved by a factor equal to the square root of the number of measurements. If n measurements are made:

$$\left.\frac{S}{N}\right|_{\text{new}} = \sqrt{n}\left.\frac{S}{N}\right|_{\text{original}} \tag{2-1}$$

As the noise always has a finite bandwidth, each noise sample has some dependence on previous values. To realize the full advantage of averaging, allow enough time between samples to ensure that they are truly random. This waiting time between signals is inversely proportional to the system bandwidth.

Another alternative for noise removal is *low-pass filtering*. As we know, however, filters become very complicated as the noise rejection increases. Very often, the concepts of averaging and filtering are combined. Through trade-offs between filter bandwidth and sample rate, an optimum system solution can be achieved.

One thing becomes obvious at this point. Although we are trying to eliminate the noise rather than examine it, we must know the noise bandwidth and other system constraints and limitations.

2–1.2 Quasi-dc Signals

A quasi-dc signal is a normally dynamic signal that does not change during the measuring period. Two examples of this type of signal are a flat-top pulse, whose amplitude is of interest, or possibly the output of an analog multiplexer. Figure 2–2 is one representation of this type of signal. Once the signal has reached the static state, all the concepts for measuring dc signals hold. Care must be exercised, however, when applying a low-pass filter to this type of measurement. System band-

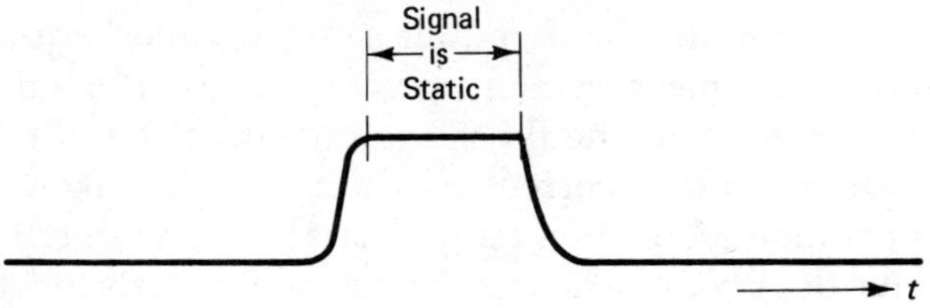

Figure 2-2. Quasi-dc Waveform

width must be adequate to pass the pulse or to allow the multiplexer to switch.

2-1.3 Dynamic Signals

The vast majority of signals we are interested in are dynamic in nature. These signals are constantly changing, often in an unpredictable fashion. As we shall see later, dynamic signals may be represented and examined either in the time domain or in the frequency domain, and we shall frequently transfer between the two to effect a proper system solution.

There are two major classes of dynamic signals:

1. *Deterministic signals* are those that have known characteristics and that can be explicitly described by mathematics.
2. *Random signals* are those whose behavior is highly unpredictable at any future time.

Let's look at each type in more detail.

Deterministic Signals

Periodic signals and transients fall under the classification of deterministic signals. *Periodic signals* are those that repeat at regular intervals, while *transients* are of an aperiodic form and decay to a zero value after a finite length of time.

Either type may be expressed in mathematical terms. Periodic functions are readily expressed as a Fourier series:

$$f(t) = \frac{a_0}{2} + \sum_{n=1}^{\infty} (a_n \cos nw_1t + b_n \sin nw_1t) \quad (2\text{–}2)$$

where:
T = the period of the signal or waveform
f_1 = the fundamental cyclic frequency = $1/T$
w_1 = the fundamental radian frequency = $2\pi f_1$
n = integer defining the order of the harmonic

The constants a_n and b_n are found by the integrals

$$a_n = \frac{2}{T} \int_{-T/2}^{T/2} f(t) \cos nw_1t \, dt, \qquad n = 0,1,2,3, \ldots \tag{2-3}$$

and:

$$b_n = \frac{2}{T} \int_{-T/2}^{T/2} f(t) \sin nw_1t \, dt, \qquad n = 1,2,3, \ldots \tag{2-4}$$

From these expressions, both an amplitude-versus-frequency and a phase-versus-frequency plot may be constructed. The amplitude plot will be a plot of $\sqrt{a_n^2 + b_n^2}$, while the phase plot will be a plot of $\tan^{-1}(-b_n/a_n)$.

Take the following exponential functions:

$$\cos nw_1t = \frac{e^{jnw_1t} + e^{-jnw_1t}}{2} \tag{2-5}$$

and:

$$\sin nw_1t = \frac{e^{jnw_1t} - e^{-jnw_1t}}{2j} \tag{2-6}$$

and substitute them into expression (2-2). You thus develop the compact form of the Fourier series:

$$f(t) = \sum_{n=-\infty}^{\infty} C_n e^{jnw_1t} \tag{2-7}$$

$$C_n = \frac{1}{T} \int_{-T/2}^{T/2} f(t) e^{-jnw_1t} dt \tag{2-8}$$

The expressions of $f(t)$ and C_n form a transform pair. The Fourier coefficient C_n is a complex number and is:

$$C_n = a_n - jb_n = \sqrt{a_n^2 + b_n^2}\, e^{j\Theta_n} \tag{2-9}$$

where

$$\Theta_n = \tan^{-1} \frac{-b_n}{a_n} \tag{2-10}$$

and the magnitude of C_n is:

$$|C_n| = \sqrt{a_n^2 + b_n^2} \tag{2-11}$$

Obviously, the plots of $|C_n|$-versus-frequency and Θ_n-versus-frequency are the same plots described by the less compact form of the Fourier series. Both plots are discrete line plots in the frequency domain.

The simplest periodic waveform is, of course, a single sinusoid. Both the time and frequency domain plots are shown in Figure 2-3. There is only one frequency domain plot, a single-amplitude spectral line at a frequency that represents the single sinusoid. Not terribly interesting, but it does point out that even the simplest of waveforms have representations in both domains.

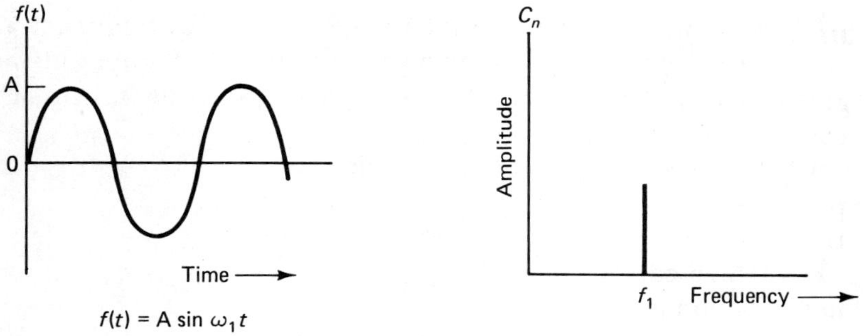

Figure 2-3. Time and Frequency Domain Representations for a Single-Frequency Signal

Figure 2-4 shows a more interesting signal waveform that consists of several harmonically time-locked sinusoidal signals. The frequency-domain magnitude plot contains a dc component and three other discrete frequency components at f_1, f_2, and f_3. Their phase relationships are shown on the lower frequency domain plot. Note that the plots are in the form of discrete line spectra, which characterize periodic waveforms.

Perhaps the most important periodic waveform, from a data acquisition point of view, is that of Figure 2-5. The rectangular pulse train shown is typical of a real, finite-width sampling function. The resultant frequency domain function is:

$$C_n = \frac{A\tau}{T} \frac{\sin n w_1 \tau/2}{n w_1 \tau/2} \qquad (2\text{–}12)$$

As it turns out, C_n is real and there is no need to plot C_n and Θ_n sepately. As Figure 2-5 indicates, the Fourier series transform produces a spectrum that is symmetrical about the $f = 0$ axis. Although the

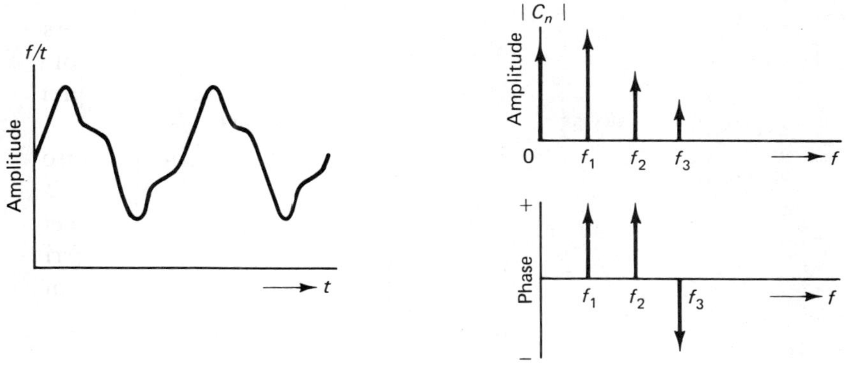

Figure 2-4. Typical Periodic Function

"negative" frequencies are fictitious, they will prove useful when the effects of sampling are considered. Should the pulse train repetition rate increase (T decreases), the number of spectral lines decreases. As the pulse width (τ) decreases, the zero crossings move out, thus extending the frequency content over a wider frequency range.

Figure 2-6 carries this concept to its limit. In this case, the time-domain function is a sequence of unit impulse functions, where a unit impulse is defined as a pulse that has infinite height and zero width. The area of the pulse, however, is unity. With an infinitely small τ, the first zero crossings in the frequency domain are at $\pm\infty$. The spectral lines are of equal amplitude as the energy is distributed evenly across the spectrum. An impulse pulse train in the time domain is then represented by an impulse spectra train in the frequency domain. This characteristic will be utilized later when explaining the effects of narrow pulse sampling.

As we know, the transient signal is an *aperiodic* signal that reduces to zero after some finite interval of time. If we consider the transform pair of (2-7) and (2-8) and allow T to approach infinity (only one pulse or transient, ever), then the fundamental frequency, ω_1, will approach zero and a new transform pair is created:

$$f(t) = \frac{1}{2\pi} \int_{-\infty}^{\infty} F(\omega)e^{j\omega t}d\omega \qquad (2\text{-}13)$$

and:

$$F(\omega) = \int_{-\infty}^{\infty} f(t)e^{-j\omega t}dt \qquad (2\text{-}14)$$

Now as a Fourier integral representation, the ability to transfer

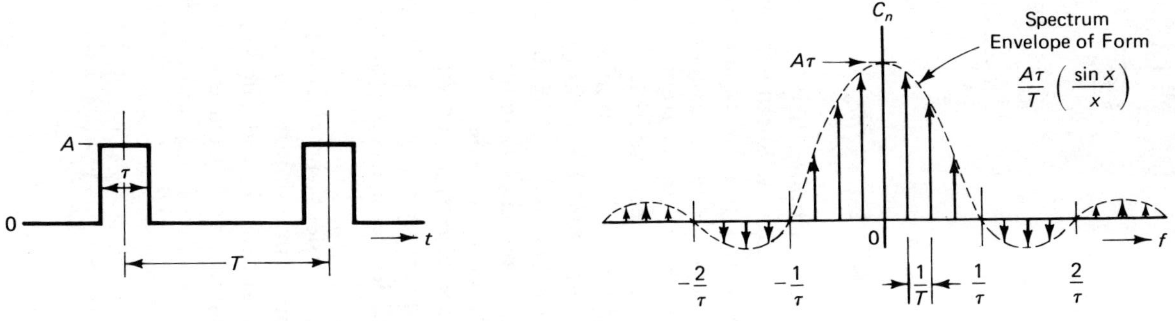

Figure 2-5. Rectangular Pulse Train with Frequency Spectrum

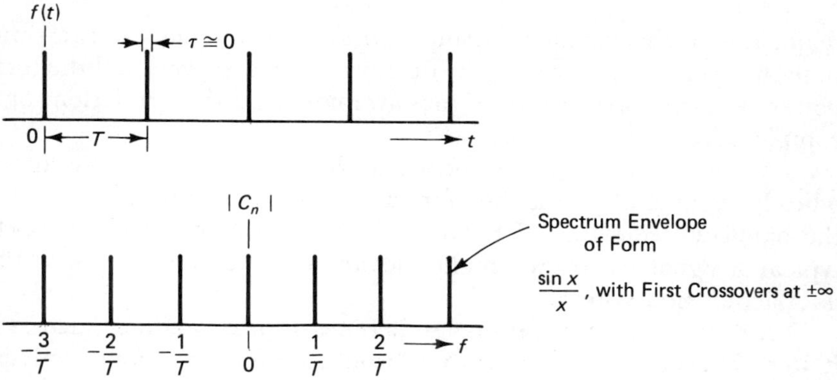

Figure 2-6. Impulse Function in Both Time and Frequency Domains

back and forth from the two domains is once again preserved. An example of the application of the Fourier integral is seen in Figure 2-7. The time domain function is a single square pulse centered about the y-axis. Although the result is still of the form $(\sin x)/x$, something interesting has occurred. The plot is not a series of discrete line spectra as we found with the pulse train of Figure 2-5. As T has gone to infinity, $1/T$ goes to zero, and all possible line spectra are present. The result is a closed spectrum, with all frequencies contained. Once again, the "negative" frequencies are produced by the transformation.

The reverse situation occurs with a "box" filter in the frequency domain, where the frequency spectrum looks the same as the single-pulse time function of Figure 2-7. The resultant time plot is, of course, of the form $(\sin x)/x$.

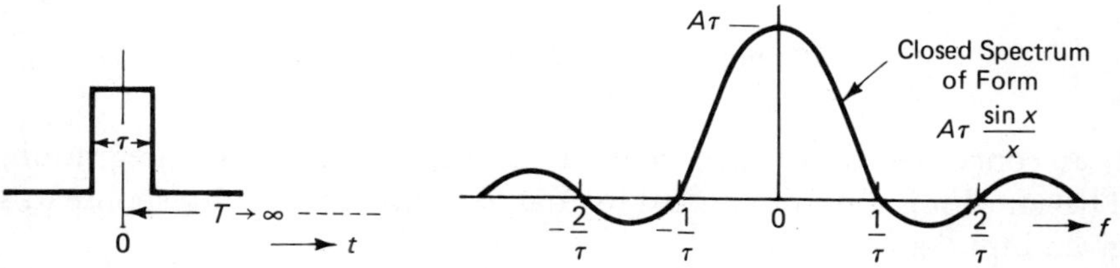

Figure 2-7. Single Pulse with Frequency Spectrum

Random Signals

Unlike deterministic signals, the exact value of a *random* signal may not be predicted in advance. Since most signals we must deal with fall into this category, the concepts of probability and statistics play a major role in their analysis. Using statistical techniques, we can compute the average or dc value, the dc power and ac power, and the total power. We can also perform signal averaging, signal correlation, or a complete frequency analysis.

As the random signal is not repeatable, its time function is aperiodic. Its frequency plot is therefore a "closed" spectrum, as shown for the band-limited signal of Figure 2-8. The spectrum shown is fairly typical of signal spectra we might encounter in the everyday use of the data acquisition system.

A deterministic signal may sometimes appear random, such as if it were buried in noise. First, you would perform random analysis, utilizing such techniques as signal averaging or correlation, to extract the deterministic signal. When that process is complete, the signal can then be analyzed by normal deterministic analysis.

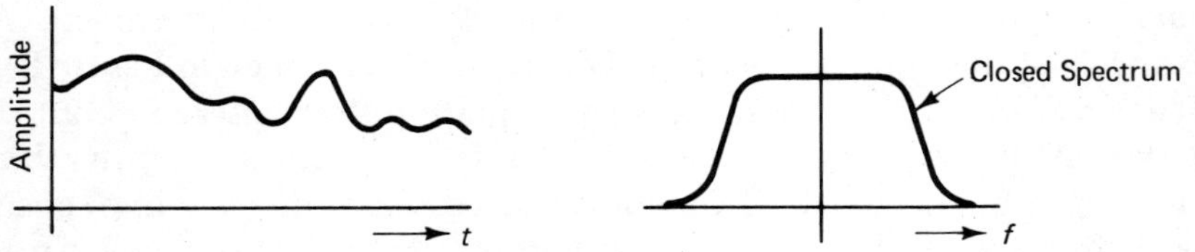

Figure 2-8. Band-Limited Nonperiodic Signal

2-1.4 Convolution

One other concept is important to a basic understanding of sampling: the principal of convolution. The time domain convolution integral is expressed as:

$$g(\tau) = \int_{-\infty}^{\infty} f(t)h(\tau - t)dt \qquad (2\text{-}15)$$

It is concerned with displacement, multiplication, and integration. The concept is most often used to analyze a circuit output response to a given input stimuli.

In the example found in Figure 2-9, the square-pulse input, $f(t)$, is applied to a simple RC circuit whose impulse function is the reverse exponential, $h(t)$. The convolution integral of (2-15) could also be written as:

$$g(\tau) = \int_{-\infty}^{\infty} f(t)h(-t + \tau)dt \qquad (2\text{-}16)$$

The functions $h(t)$ and $h(-t)$ can be seen in Figures 2-9b and 2-9c. We see that $h(t)$ is obtained by folding $h(t)$ about the zero-time axis. The

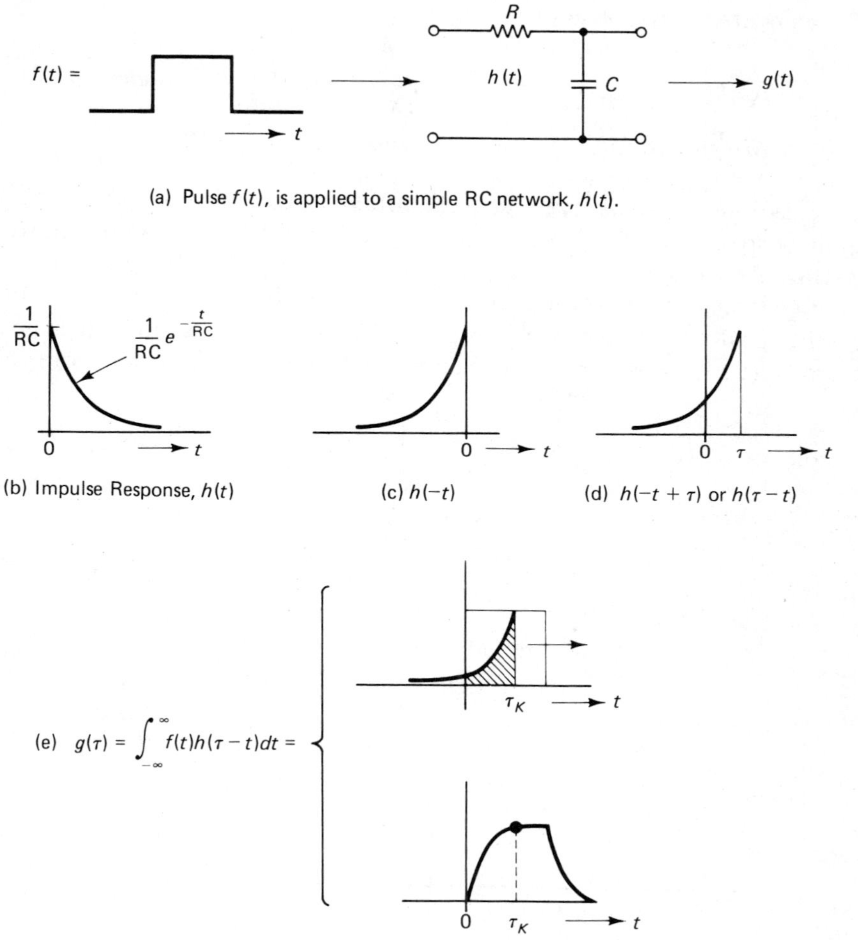

Figure 2-9. An Example of Convolution

function $h(-t + \tau)$ represents the function $h(-t)$ shifted by τ-seconds along the time axis, as seen in Figure 2-9d. The resulting effect of convolution (see Figure 2-9e) is to move the reverse impulse response curve, $h(-t + \tau)$, across the time function, $f(t)$, while evaluating the shaded area for each new value of τ. The net result is the lower part of Figure 2-9e, the summation of all the shaded areas.

It is also important to realize that convolution in the time domain is equivalent to multiplication in the frequency domain, and:

$$\underset{\textit{Convolution}}{\int_{-\infty}^{\infty} f(t)h(\tau - t)\,dt} \quad \leftarrow \; \rightarrow \quad \underset{\textit{Multiplication}}{F(\omega)H(\omega)} \tag{2-17}$$

The converse is also true, as:

$$f(t)h(t) \quad \leftarrow \; \rightarrow \quad \frac{1}{2\pi}\int_{-\infty}^{\infty} F(\omega)H(W - \omega)d\omega$$

Multiplication *Convolution* (2–18)

As one last example, let's examine the convolution of the frequency spectrum of Figure 2-10a with the impulse function of Figure 2-10b. As we see, both frequency spectra are perfectly symmetrical about zero, and, therefore, $H(W - \omega)$ is exactly $H(\omega)$. Evaluation of the integral may be easily accomplished visually. As the impulse spectrum, $H(\omega)$, passes across, or is convolved with, the $f(\omega)$ spectrum, a

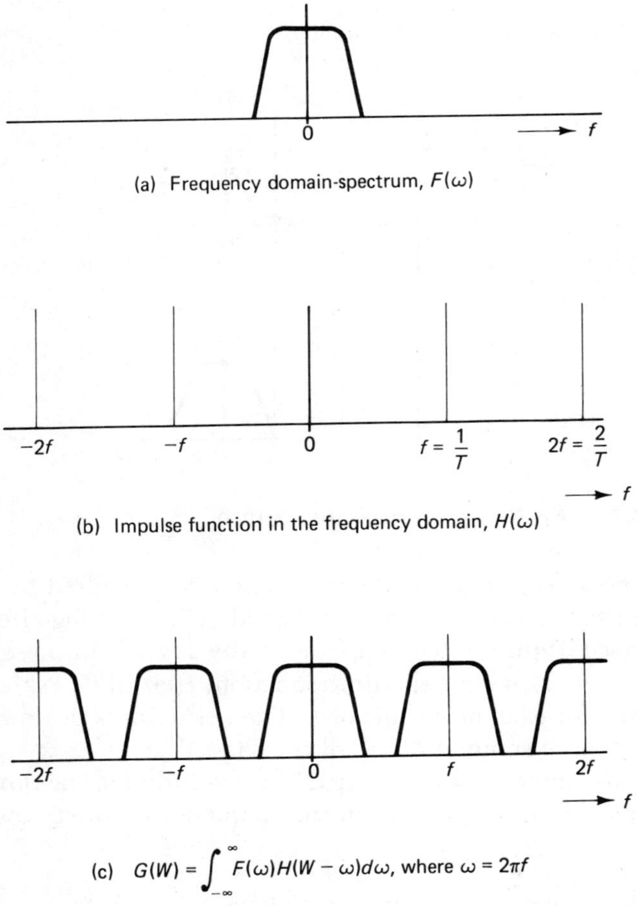

(a) Frequency domain-spectrum, $F(\omega)$

(b) Impulse function in the frequency domain, $H(\omega)$

(c) $G(W) = \int_{-\infty}^{\infty} F(\omega)H(W - \omega)d\omega$, where $\omega = 2\pi f$

Figure 2-10. Convolution of a Frequency Spectrum with a Unit Impulse Train completely new spectrum is imprinted, symmetrically about each impulse.

2-1.5 Important Things to Remember

Before proceeding to the concepts of sampling, note the following points:

1. There are three basic kinds of signals: dc, quasi-dc, and dynamic.
2. Dynamic signals have both time and frequency domain representations.
3. We may readily transfer from one domain representation to the other and vice versa.
4. Multiplication in the time domain is equivalent to convolution in the frequency domain and vice versa.

2-2 THE SAMPLING FUNCTION

In Section 2-0, we defined sampling as the act of measuring a continuous signal at discrete time intervals. So we could consider the sampler as a simple switch. When the switch is closed, it is equivalent to multiplying the input signal by unity for the duration of the closure period. The results of sampling with a unit-amplitude impulse train is shown in Figure 2-11, where the signal, $f(t)$, is sampled by $h(t)$. The output is the impulse train, $f^*(t)$, whose amplitudes trace out the envelope of the original signal, $f(t)$.

In actual practice, hardware constraints place severe limitations on our ability to achieve a true impulse sample function. For ease of understanding, however, we will utilize the impulse function to explain the basic sampling concepts, describing the added effect of finite-width sampling near the end of the chapter.

So the sampling function consists of multiplication in the time domain, as defined by the expression:

$$f^*(t) = f(t)h(t) \qquad (2\text{-}19)$$

We also know that multiplication in the time domain is convolution in the frequency domain. The effect of sampling in both domains is shown in Figure 2-12. From the information presented in the previous sections, we should not be surprised at the results. As seen in Figure 2-12a, the random time function translates to a closed spec-

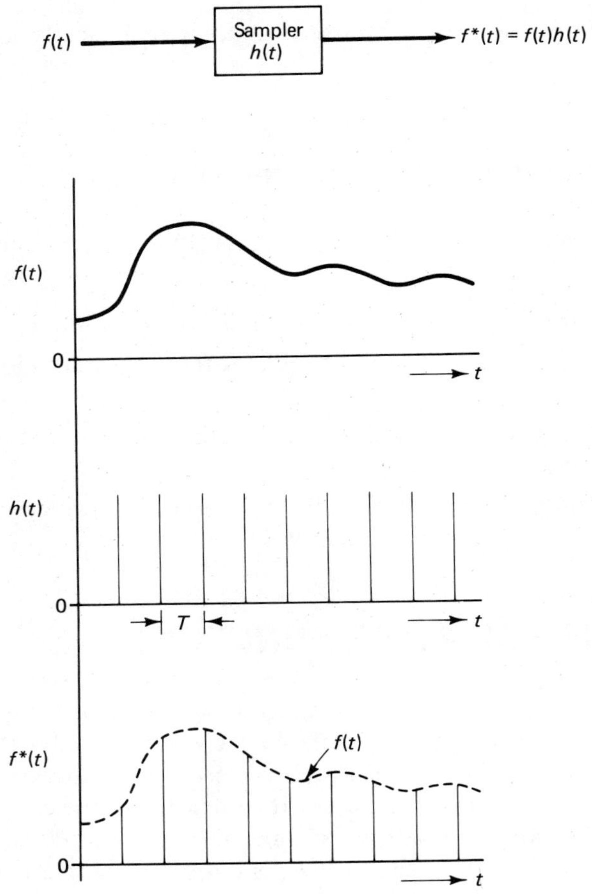

Figure 2-11. The Impulse Sampler

trum in the frequency domain, while, as Figure 2-12b shows, the impulse train is an impulse train in both domains. Finally, in 2-12c, we find that the amplitude-modulated sampled time function, $f^*(t)$, appears as a series of spectra, as the frequency spectrum, $F(\omega)$, is convolved with the impulse function, $H(\omega)$. Each of the spectra of $F^*(\omega)$ is centered about multiples of the sample frequency, $f_S = 1/T$. Note that these multiple spectra *are* present in the sampler output and can become a problem later if complete signal recovery and reconstruction are required.

We learned in Section 2-1.3 that periodic functions in the time domain become sampled spectra (Figures 2-4 and 2-5) in the frequency domain. Now we find that sampled functions in the time domain translate to periodic spectra in the frequency domain.

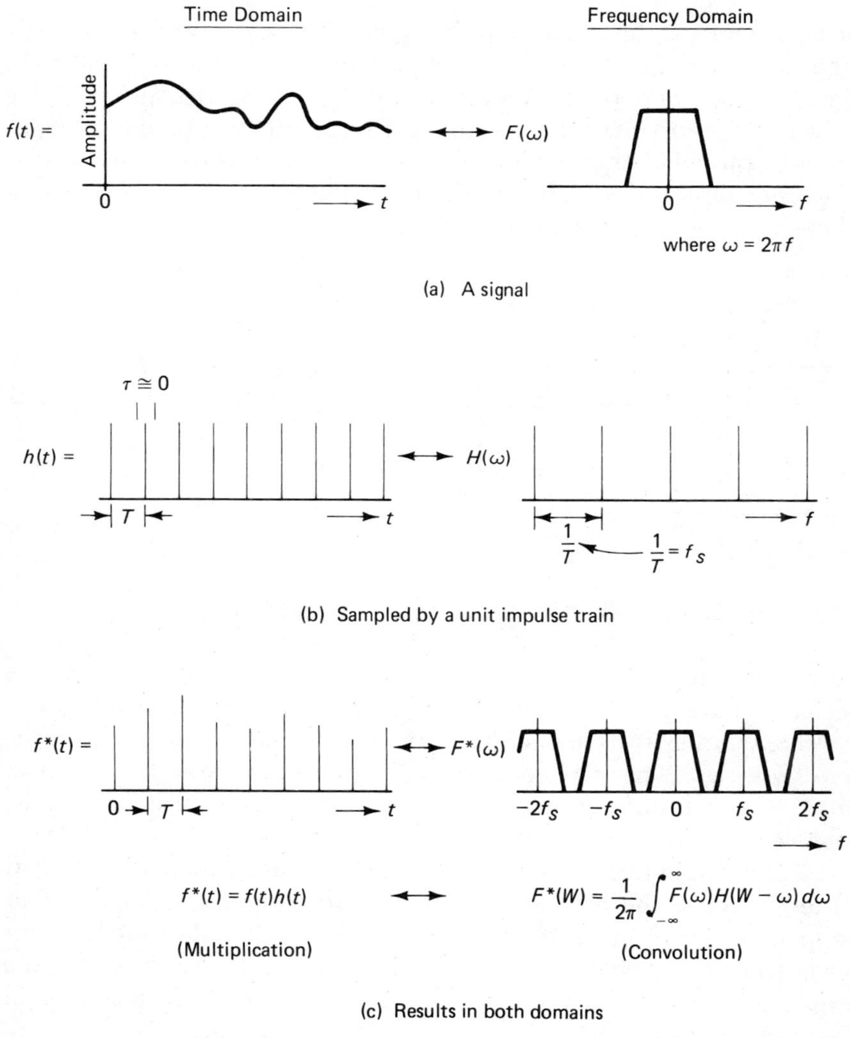

Figure 2-12. Sampling in the Time and Frequency Domains

2-3 THE SAMPLING THEOREM

Our discussion of sampling would not be complete without reference to the *sampling theorem*, which states that:

> signals with a finite bandwidth of *f*-hertz can be completely described by sampling the time signals at instants separated by $T = 1/2f$ seconds.

In simple terms, the signal must be sampled at a rate at least twice as high as the highest frequency in the spectrum. This concept is shown in Figure 2-13, where the sinusoid is sampled at its positive and negative peaks. The theory then assumes that the resultant impulse sample train could be passed through an ideal "box" filter and the original sinusoid *frequency* reproduced. The amplitude of the filter output depends on where the original sinusoid is sampled.

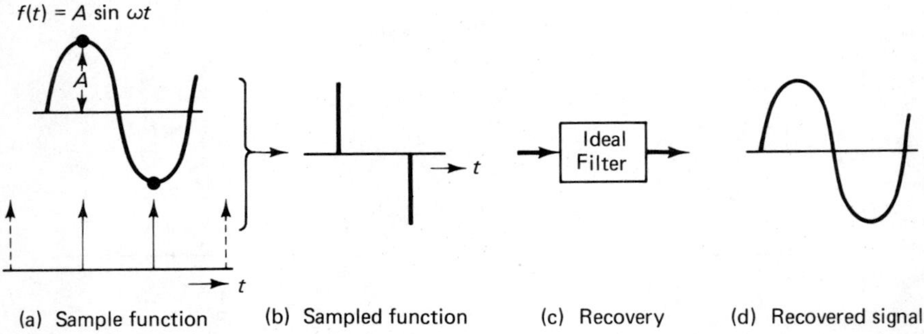

Figure 2-13. Basics of the Sampling Theorem

Since the sampling theorem defines the limit for an absolutely perfect situation, it has a number of serious problems. First, suppose we were so unlucky as to choose sample points that fall precisely on the zero crossings. This case, shown in Figure 2-14a, would, of course, produce zero output from the sampler. While the case is trivial, the problem is fundamental.

The second difficulty arises in trying to build a filter that cuts off perfectly above the highest frequency, *f*. In contrast, consider what happens if the reconstruction filter is a simple digital-to-analog converter (zero-order hold). The result, shown in Figure 2-14b, is that the square wave has many more frequencies contained than the original single sine wave.

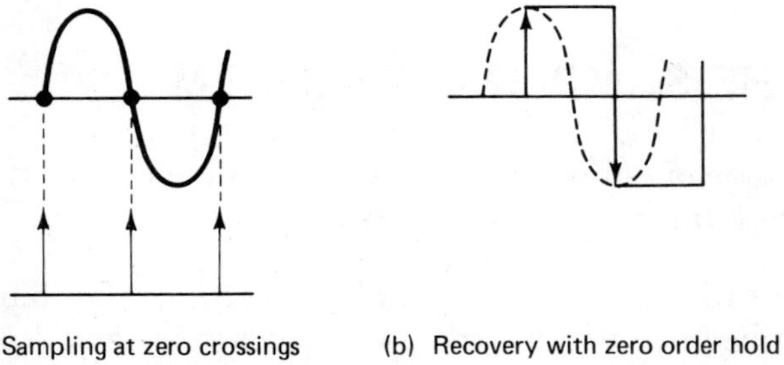

Figure 2-14. Obvious Difficulties with the Sampling Theorem

The last, and perhaps most serious, problem is that we tend to become attached to the "highest frequency of interest" and often also consider it to be f, the "highest frequency contained." In actuality, no spectrum exists that cuts off perfectly at f, and substantial energy exists in the frequencies beyond it. This brings us to the concept of *aliasing*.

2-4 ALIASING

Webster's dictionary defines the word alias as "an assumed name" or "to be called by another name." The concept of aliasing, in a sampled data system, means the same thing when referred to frequencies. When a given frequency is sampled at too low a rate, it appears as a totally different lower frequency at the output of the sampler. So we say that the resultant frequency is the "alias" of the original.

An example of aliasing in the time domain is shown in Figure 2-15. The actual signal is being sampled at a rate that is somewhat less than one sample per cycle. The resultant frequency is in the vicinity of one-third the original signal. If we were able to accurately read the waveforms of Figure 2-15, we would find:

$$f_{alias} = f_{actual} - f_{sample} \qquad (2\text{–}19)$$

In fact, f_{alias} is always the difference between the actual signal and some harmonic of the sample frequency (f_S). Consider an example where $f_{sample}(f_S) = 100$ kHz and the signal contains the frequencies 75 kHz, 125 kHz, and 175 kHz. The sampler results are:

f	*Formula*	*Result*
75 kHz	$f_S - f$	100 − 75 = 25 kHz
125 kHz	$f - f_S$	125 − 100 = 25 kHz
175 kHz	$2f_S - f$	200 − 175 = 25 kHz

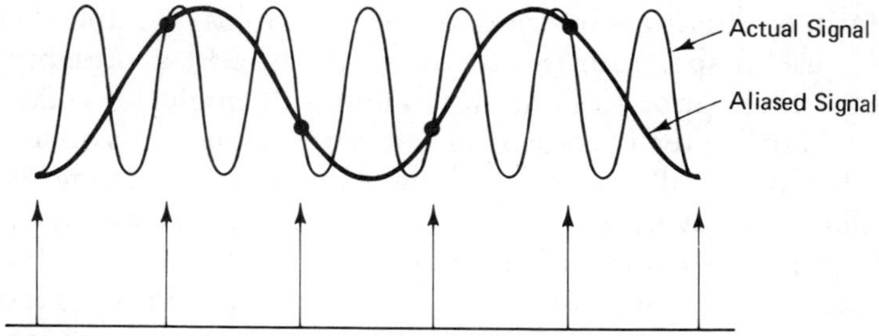

Figure 2-15. Aliasing in the Time Domain

As we see, all three frequencies "look" like 25 kHz as they are aliased.

Other interesting observations can be made if we examine the effects of aliasing in the frequency domain. In Section 2-2, we discovered that, when we sampled a signal in the time domain, it appeared as a system of multiple spectra in the frequency domain with each spectrum centered about a multiple of the sample frequency, f_S (Figure 2-12).

Consider what happens if we reduce the sample rate. In the time domain, the sample pulses spread apart. In the frequency domain, however, the spectra move closer together. Eventually, if we continue to reduce the sample rate, the spectra touch and finally overlap. Once the spectra have overlapped, aliasing has occurred. Such a situation is found in Figure 2-16. An important item to note is that the aliasing occurs about a point that is equal to one-half the sample frequency, or $f_S/2$. This point is often referred to as the *folding frequency* because all frequencies above $f_S/2$ are "folded" back into the spectrum below it. As an interesting side note, consider the rectangular spectra of Figure 2-17. The original spectrum and the first harmonic spectrum can be butted tightly together at the very point of folding, $f_S/2$.

The following expression may be written:

$$f_S - f_c = f_c, \text{ where } f_c = \text{ the cut-off frequency}$$

or:
$$f_S = 2f_c$$

and:
$$f_c = f_S/2$$

This pictorial of the sampling theorem shows that, in the perfect situation, the sample frequency must be equal to or greater than twice the highest frequency contained in the spectrum. It also clearly indicates that aliasing or folding occurs about $f_S/2$.

The obvious cure for aliasing is, of course, to sample at a high enough rate. Yet doing so is not always feasible due to hardware constraints. Also recognize that signals usually contain noise, which is much wider in its bandwidth than the frequencies of interest. Unless some form of filtering is interjected, the noise folds over and over, back into the useful spectrum. So then, we might add a presampling or "aliasing" filter prior to sampling. The result might look like Figure 2-18. A "box" filter is chosen, and it is therefore impossible to construct. But it does illustrate the point. The spectra no longer overlap.

Like so many solutions, an aliasing filter is not a panacea, and certain problems come with it. First, complex filters introduce time delay, which must be considered if they are placed in the middle of a process control feedback system. They are also expensive. If a multichannel system requires fast analog multiplexing, each channel

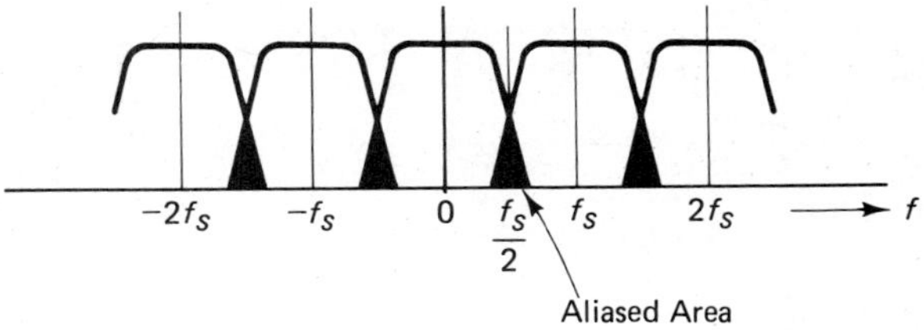

Figure 2-16. Aliasing in the Frequency Domain

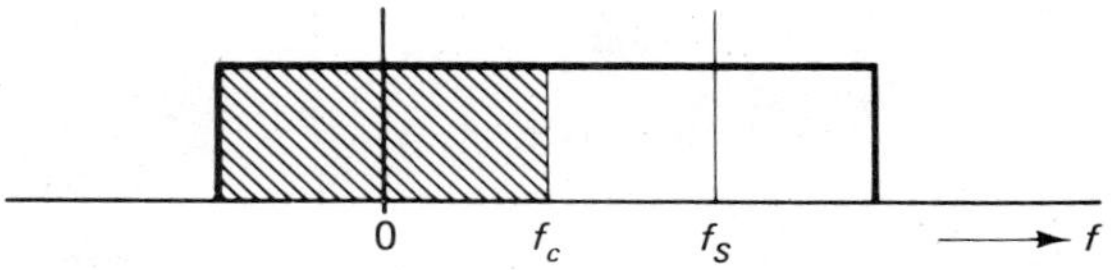

Figure 2-17. Sample Theorem with Ideal Spectrum

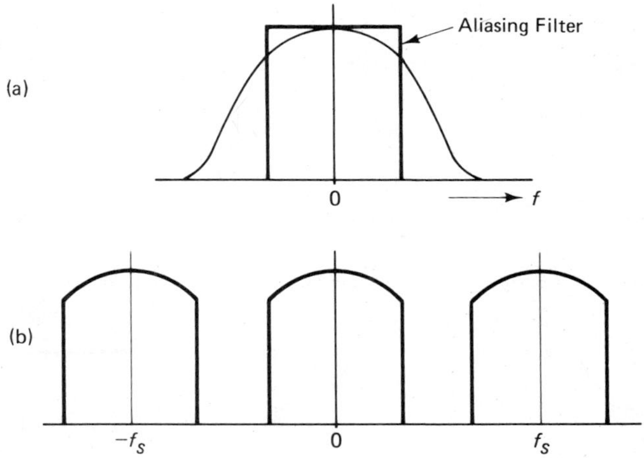

Figure 2-18. Removal of Aliasing with a Presampling or Aliasing Filter (a) and Result (b)

must have a separate filter prior to the multiplexer. The cost per filter must then be multiplied by the number of channels. The solution to a given problem is usually a combination of sample rate selection and presample filtering.

Such a situation is found in Figure 2-19. The given signal spectrum has a peak or hump at frequencies well beyond the useful range. Sampling at a rate commensurate with the useful portion of the spectrum will fold the unwanted peak back into the middle of the information to be observed, as shown in Figure 2-19b. The problem could be solved by substantially increasing the sample rate, as Figure 2-19c

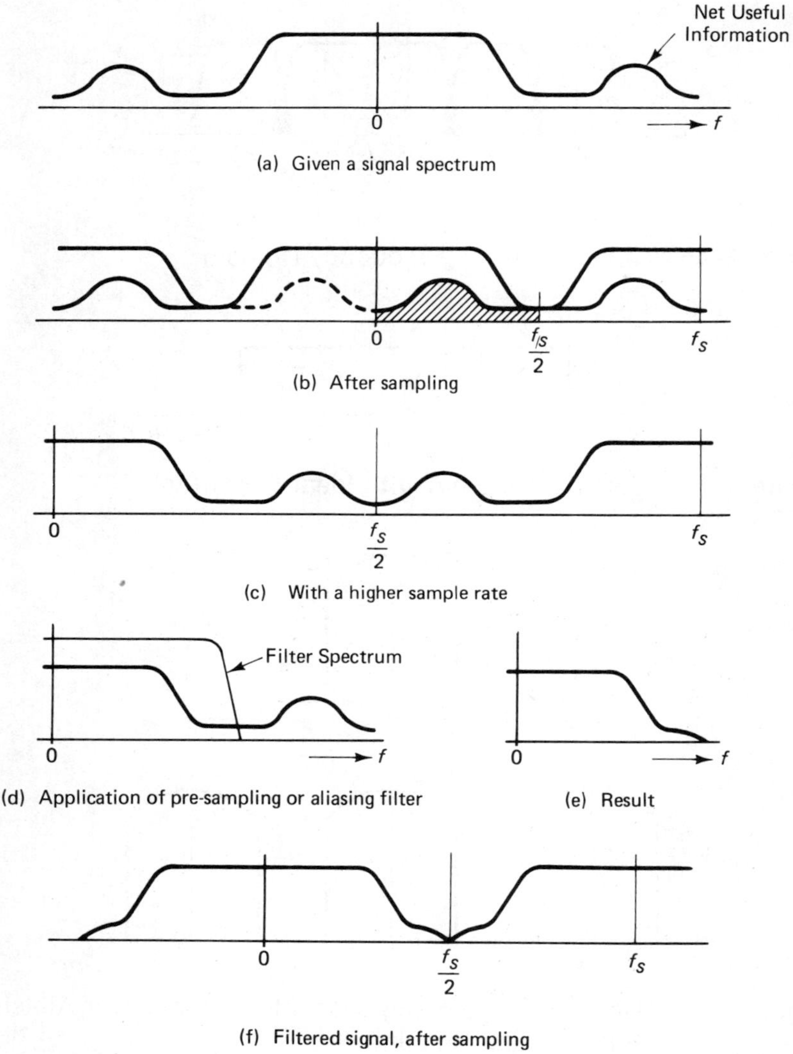

Figure 2-19. Aliasing Example

shows. Suppose, however, we utilize a reasonable presampling filter like that of Figure 2-19d. In the resultant spectrum, Figure 2-19e, the peak is removed and a lower sample rate may be utilized. Figure 2-19f shows a satisfactory solution for the problem with a final sample rate that is not much greater than that first tried.

Another alternative might be a post-sampling filter as found in Figure 2-20. This solution is not often applied because finding an adequate filter becomes very difficult. The filter must remove the aliasing

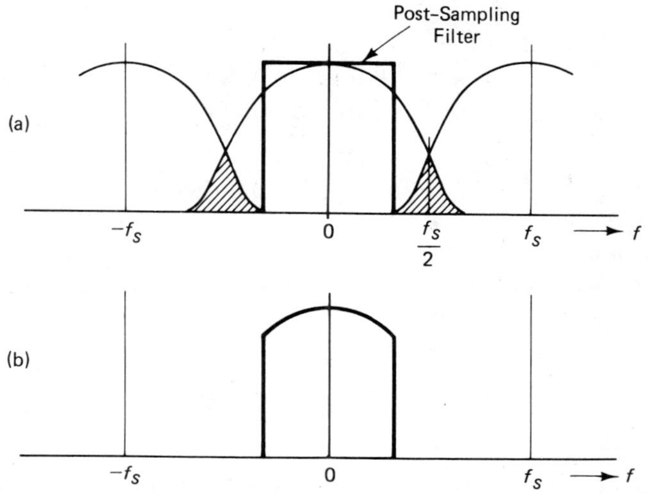

Figure 2-20. Removal of Aliasing with a Post-Sampling Filter (a) and Result (b)

area without causing degradation to the wanted portion of the spectrum. It also precludes the salvaging of any information that lies in the folded area.

Most systems are based on a compromise of sample rate with presample filtering and, as we will discover in the next chapter, the signal recovery or reconstruction technique. One thing to keep in mind is that you can never solve the problem completely. The noise bandwidth almost always exceeds the wanted signal bandwidth and so some is always folded back. The problem becomes not whether folding should occur but rather how much of it can be tolerated.

2-5 BAND-PASS SAMPLING

So far, our discussion has considered video-type signals where the spectrum starts at dc, or zero frequency, and goes out to some frequency where it falls to zero. Another class of signals has no dc component but rather starts at some low frequency, f_L and goes to some high frequency, f_H. An example of such a signal is found in Figure 2-21a. Just as we would expect, when the signal is sampled, multiple band-pass spectra appear, centered about multiples of the sample frequency. There is a *band-pass sampling theorem* that applies to this situation, as:

$$\frac{2f_L}{K} > f_S > \frac{2f_H}{K+1} \qquad (2\text{–}20)$$

where

$$f_L = \text{low 3-dB frequency}$$
$$f_H = \text{high 3-dB frequency}$$
$$f_S = \text{sample frequency}$$
$$K_{\max} < \frac{f_L}{f_H - f_L} = \frac{f_L}{\text{Bandwidth}}$$

The lowest sample frequency, f_S, is found by using the largest *integer, K*, which gives a valid equality for (2–20). It can be proven that at $K_{\max}$:

$$f_S = \frac{2F_L}{K} = \frac{2f_H}{K+1} = 2 \times \text{Bandwidth}$$

which is a corollary to the basic sampling theorem.

Let's consider what happens if the sample rate decreases. Looking at Figure 2–22a, we see that folding occurs when the spectra first touch, where $f_S - f_H = f_H$, or $f_S = 2f_H$. Now look at Figure 2–22b and observe what happens. The folding *stops* when $f_S - f_L$ equals f_L and the sample rate is $2f_L$. This rate is lower than the rate where folding

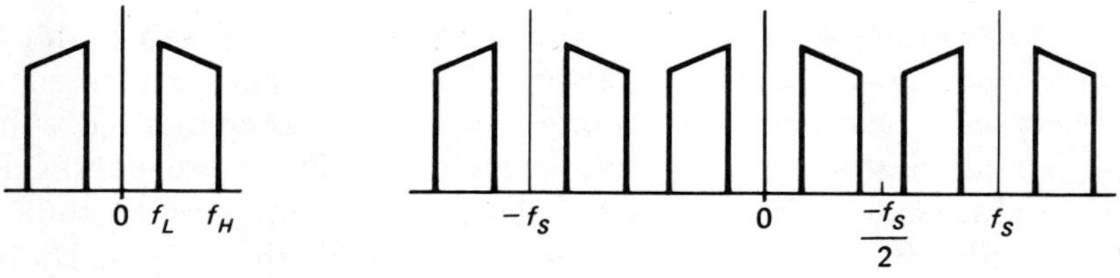

(a) Bandpass signal (b) After sampling

Figure 2–21. Band-Pass Signal Properly Sampled

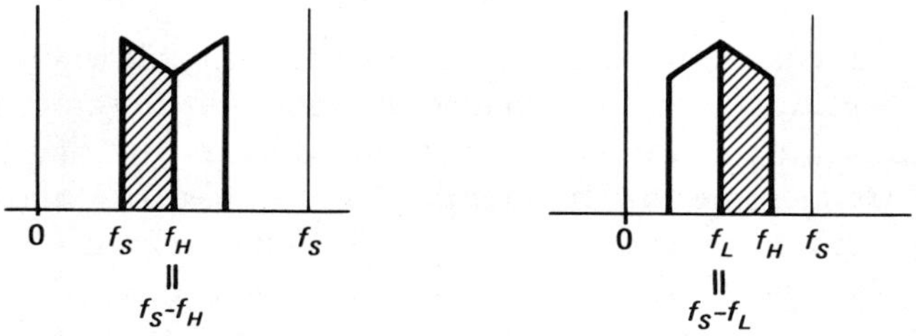

(a) Folding starts, $f_S = 2f_H$ (b) Folding stops, $f_S = 2f_L$

Figure 2–22. Band-Pass Sampling with Reduced Rates

started by two times the bandwidth. Should the bandwidth be small with respect to f_L, folding occurs in many areas and then stops as the sample frequency is reduced, while the unwanted spectra continue to fill in the gap between the zero axis and f_L.

Consider a problem where:

$$f_H = 250 \text{ kHz} \qquad f_L\ 210 \text{ kHz}$$

and therefore:

$$\text{Bandwidth } f_H - f_L = 40 \text{ kHz}$$

K_{max} is then:

$$K_{max} = \frac{210}{40} \cong 5 \qquad (K \textit{ must be an integer})$$

Then, using (2-20), f_S is:

$$\frac{420}{5} > f_S > \frac{500}{6}$$

or: $$84 \text{ kHz} > f_S > 83.3 \text{ kHz } (\textit{choose } 83.5 \textit{ kHz})$$

To analyze this problem further, observe the chart in Figure 2-23. The values of K from $K = 5$ to $K = 0$ have been plotted. You should note that the spectrum plots always lie totally on a side of one of the triangles. At the lowest sample rate, $K = 5$, the spectrum nearly fills the side, while at the high rate, $K = 0$, it occupies only a small portion of the $K = 0$ slope. Should any of the spectrum plots cross over and occupy space on more than one side of a triangle, it is a sign that aliasing is occurring. As the sample rate gets lower, joining up the spectra and ensuring a result that is free of folding both become more difficult. Sample rates higher than 600 kHz simply reduce the size of the spectrum plot relative to the triangle side and move it down the slope towards zero. Also note that we have been working with infinitely steep-sided spectra, while more realistic ones are rounded and tail off as they move towards zero, from both f_L and f_H. The concept is valid, however, and it may be utilized if you understand the content of the input signal and its resulting frequency domain spectrum.

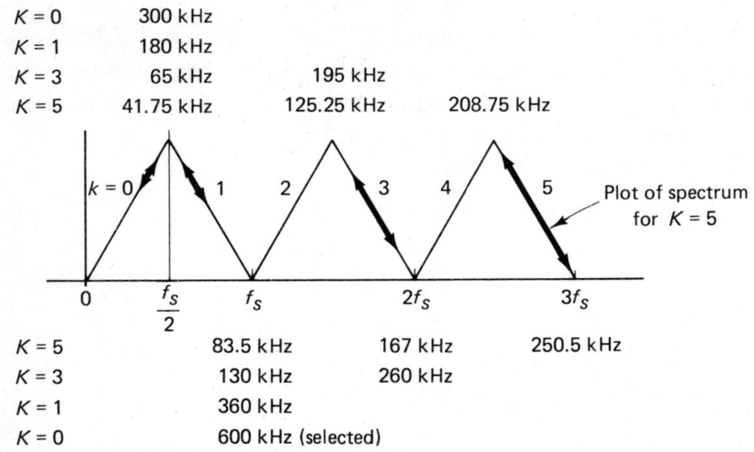

Figure 2-23. Example of Band-Pass Sampling

2-6 FINITE-WIDTH SAMPLING AND APERTURE TIME

Unfortunately, impulse sampling is not realizable in a real system environment. Regardless of how narrow the effective pulse width or sampling window is with respect to the frequency component being observed, it is always finite. The effect in the frequency domain is shown in Figure 2-24. As the sample pulses start to have finite width in the time domain, the frequency domain plot starts to develop $(\sin x)/x$ characteristics, and the resultant multiple spectrum amplitudes change correspondingly. Obviously this produces an effective filtering action that acts on unwanted spectra. Before we get too enthusiastic and decide to always sample with wide pulses, let's recognize that the same filtering action deteriorates the original spectrum as well. This point brings us to the subject of aperture time.

Aperture time refers to the width of the sampling window or of the actual time the sampler requires to obtain an accurate reading of the analog input signal. Of concern is the fact that the signal to be measured is changing while the system is acquiring a reading. If the sample function could be truly an impulse and infinitely narrow, the sampler output would be precisely the input value at the point of sampling.

Consider the sine wave of Figure 2-25. If this waveform is to be measured accurately at any point, the sampler must successfully handle the point of the greatest rate of change or the zero crossing. The slope at that point is the derivative of $V(t)$ evaluated at the zero crossing, or:

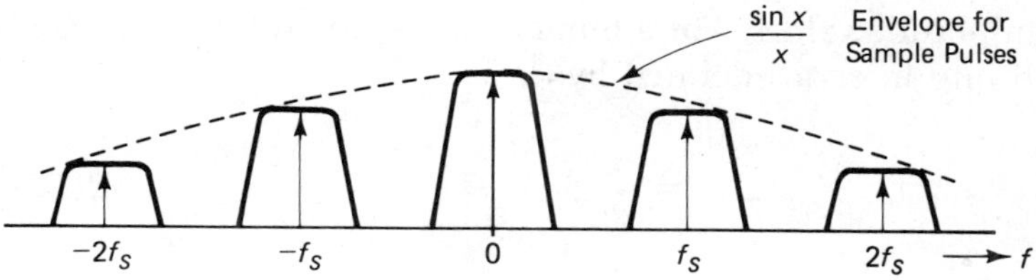

Figure 2-24. Effects of Finite-Width Sampling

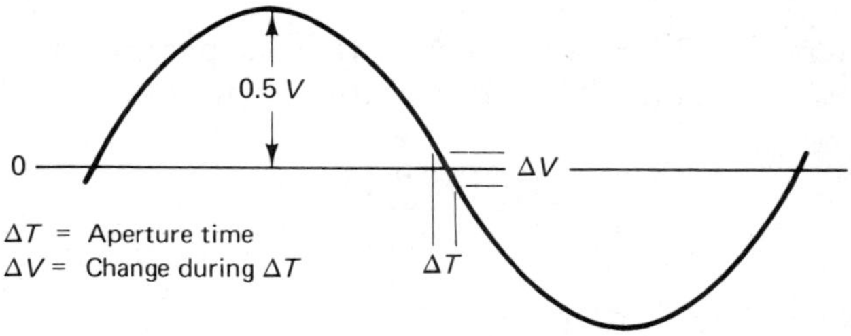

Figure 2-25. Sine Wave, $V(t) = 0.5V \sin \omega t$

$$\left.\frac{dV(t)}{dt}\right|_{max} = 0.5V\omega \cos \omega t\Big|_{\text{at zero}}$$

As $\cos \omega t = 1$ at zero, and $\omega = 2\pi f$, the result is:

$$\left.\frac{dV(t)}{dt}\right|_{max} = 0.5V\omega = (0.5V)(2\pi f) = \pi f V \text{ Volts/sec}$$

Referring to the same figure, we can express the maximum slope with reference to the aperture time by:

$$\frac{\Delta V}{\Delta T} = \pi f V \qquad (2\text{–}21)$$

where $\Delta T = \text{aperture time} = \tau_\alpha = \frac{\Delta V}{V} \times \frac{1}{\pi f}$

To make (2–21) track with the aperture time formulas that are common to the data acquisition industry, we have to jump ahead and introduce the concept of quantization (a more detailed discussion will be found in Chapter 3). In the data acquisition system, the output of the sampler is further translated into the digital world by means of a quantizer. The quantizer allows the sampled signal to have only discrete amplitude values. For a binary system, the full-scale range is segmented into intervals defined by:

$$\Delta V = \frac{V}{2^n} \tag{2-22}$$

where n = the number of bits
V = full-scale voltage

With reference to (2–21), we recognize that the parameter V is directly related to the ΔV of (2–22) and, substituting, we find that:

$$\begin{aligned} \tau_\alpha &= \frac{V/2^n}{V} \times \frac{1}{\pi f} \\ &= \frac{1}{2^n \pi f} \end{aligned} \tag{2-23}$$

Given the aperture time and number of bits, we find the maximum frequency that can be sampled to the accuracy implied:

$$f = \frac{1}{2^n \pi \tau_\alpha} \tag{2-24}$$

These expressions are predicated on the input moving a full quantum level, or the least significant bit (LSB), during the aperture window. Should the system be specified on the basis of ±½LSB, the expressions become:

$$\tau_\alpha = \frac{1}{2^{n+1} \pi f} \tag{2-25}$$

and:

$$f = \frac{1}{2^{n+1} \pi \tau_\alpha} \tag{2-26}$$

Some waveforms are more readily examined directly in terms of their rate of change rather than of their frequency content. Take, for instance, the signal of Figure 2–26. The relationship here is:

$$\text{Slope} = \frac{dv}{dt}$$

$$= \frac{\text{Quantum interval}}{\text{Aperture time}}$$

$$= \frac{1/2^n}{\tau_\alpha}$$

so:

$$\tau_\alpha = \frac{1}{2^n \dfrac{dv}{dt}} \qquad (2\text{–}27)$$

For situations demanding ½LSB, the expression becomes:

$$\tau_\alpha = \frac{1}{2^{n+1} \dfrac{dv}{dt}} \qquad (2\text{–}28)$$

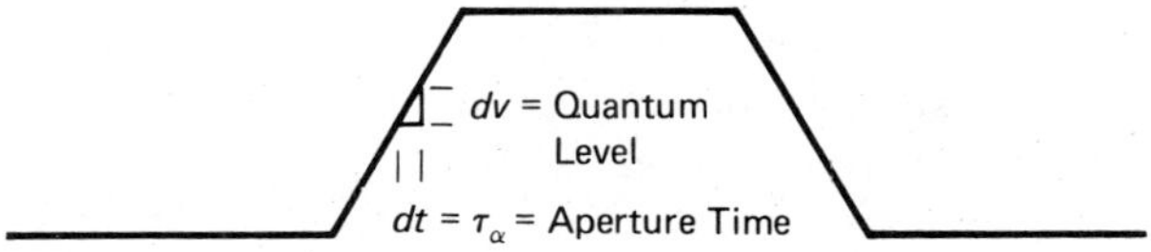

(a) Aperture time

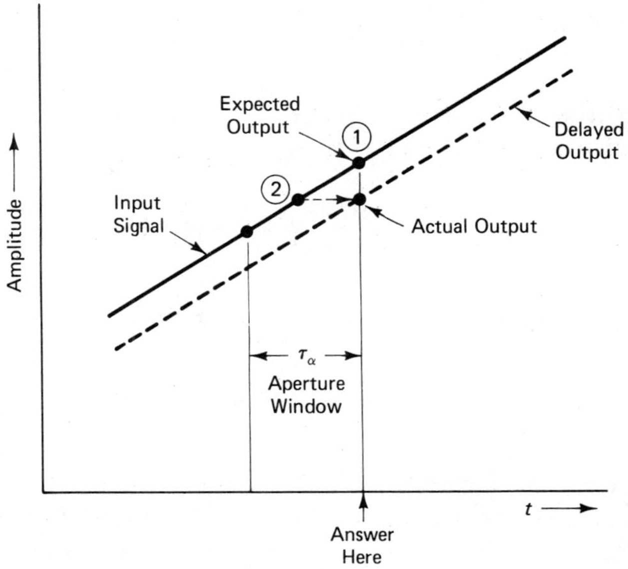

(b) Aperture uncertainty

Figure 2–26. Trapezoidal Waveform

In actuality, the aperture time window consists of two parts. The first is the effective sample pulse width, as it manifests itself at the measurement point. The second is the uncertainty of the measurement window positioning from sample point to sample point. This latter effect, called *aperture uncertainty*, manifests itself as "jitter" on the sampling waveform. In a properly designed system, the aperture uncertainty is much smaller than the basic aperture time, with the difference ranging from three times up to an order of magnitude. If the aperture time is small compared to the period of the highest frequency being sampled—that is, $\tau_\alpha < 1/10f$—the waveform is simply displaced in time after sampling, and aperture uncertainty may be used in place of aperture time. An example of aperture uncertainty can be found for the trapezoidal waveform in Figure 2-26b. The answer expected at the end of the aperture window is the value seen at point one. The actual output, however, is closer to the average signal value of point two, as the signal has been moving during the window. The effect of a series of samples is to simply displace, or delay, the output in time. That this uncertainty can be significant is shown by the following example.

Given a sampled data system with:

$$\begin{aligned} \tau_\alpha &= 50 \text{ nsec} \\ \tau_u &= \text{aperture uncertainty} = 4 \text{ nsec} \\ n &= \text{number of bits} = 10 \\ \text{accuracy} &= \pm 1/2 \text{LSB} \end{aligned}$$

If we utilize the aperture time, τ_α, the highest frequency we could examine accurately at all points is:

$$f = \frac{1}{(2^{11})\pi(50 \times 10^{-9})} = 3.1 \text{ kHz}$$

Using aperture uncertainty in place of τ_α, we find:

$$f = \frac{1}{(2^{11})\pi(4 \times 10^{-9})} = 38.9 \text{ kHz}$$

an improvement of greater than 10 to 1.

This discussion considers situations where instantaneous aperture error is important. Very often, however, we are concerned with the average error introduced at the frequency of interest. In such cases, we would consider the aperture window as an aperiodic, rectangular pulse, in the time domain, and multiply its $(\sin x)/x$ frequency domain representation with the input signal spectrum. If the aperture window is too wide, the input spectrum is degraded by the (sin x)/x filtering action. For instance, if the aperture window is 1 μsec in width, 1- to 2-percent degradation of the input signal would take place around 40 kHz.

2-7 FREQUENCY RESOLUTION

Often the data acquisition system is required to determine which frequencies are present in a spectrum and how much energy each frequency contains, relative to the others. Such an application is called *spectral* or *spectrum analysis*. Since this application means the system must be able to separate one frequency from another, we are interested in *frequency resolution*. In a general sense, the frequency resolution postulate says that if a function is available for K seconds, $1/K$ cycles per second (cps) may be resolved. In other words, at least one complete cycle of the frequency must be seen. Although spectrum analysis requires adequate sample rate selection, like any other sampled data process, the resolution is a function of the viewing period and not of the sampling frequency. Take two examples:

Example 1:

Observe for:	100 sec
Sample rate:	1,000 samples/sec
Foldover frequency:	500 cps
Resolution:	1/100 cps

Example 2:

Observe for:	100 sec
Sample rate:	5,000 samples/sec
Foldover frequency:	2,500 cps
Resolution:	1/100 cps

As we see, the higher sampling rate affects folding or aliasing, but it does not affect resolution.

A factor that does have a substantial effect is the shape of the viewing window. The simplest of shapes, the rectangular window of Figure 2-27, looks like an aperiodic pulse of width K. So it translates to a (sin x)/x function in the frequency domain. The single sine wave becomes a single spectral line. Application of the window function to the single frequency is a multiplicative process. So the two functions

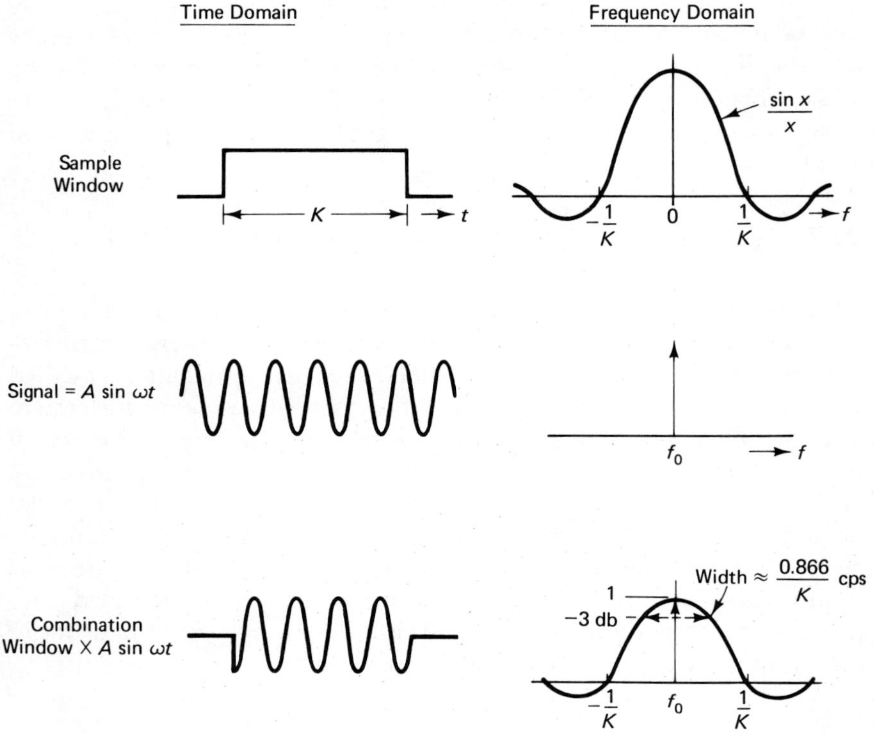

Figure 2-27. Rectangular Window with a Single Sine Wave

are convolved in the frequency domain to produce a (sin x)/x spectrum, which is centered about the f_0 spectral line. Although this function can be implemented by a simple switch, it is not an optimum window function. Figure 2-28 shows the frequency domain result for three frequencies. In this diagram, f_1 clearly stands out. The other two, f_2 and f_3, are starting to show considerable overlap. If the composite were drawn showing the effects of all the (sin x)/x tails, they would be less separable. As more and more frequencies are present, f_2 and f_3 would become more and more difficult to separate.

Fortunately, several other very tailored windows have been developed to solve this problem. The most popular of these are found

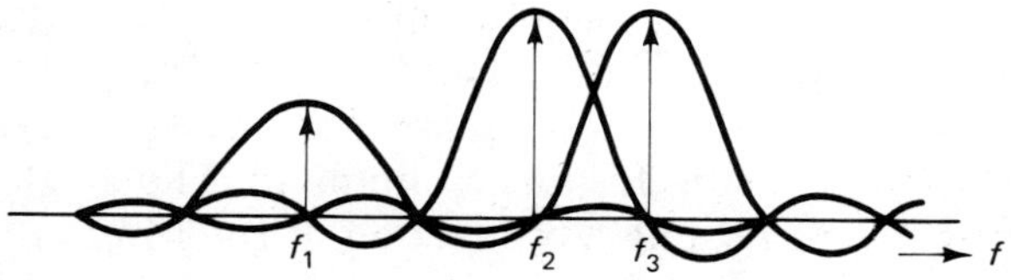

Figure 2-28. Rectangular Window with Three Frequencies

in Figure 2-29. Each has less ringing or residue in their frequency plot tails than the rectangular-(sin x)/x combination. While the separation is cleaner for these window functions, several penalties are associated with them. Besides being difficult to implement in hardware, the resolution is no longer $1/K$ but ranges from $1.5/K$ to $2.0/K$.

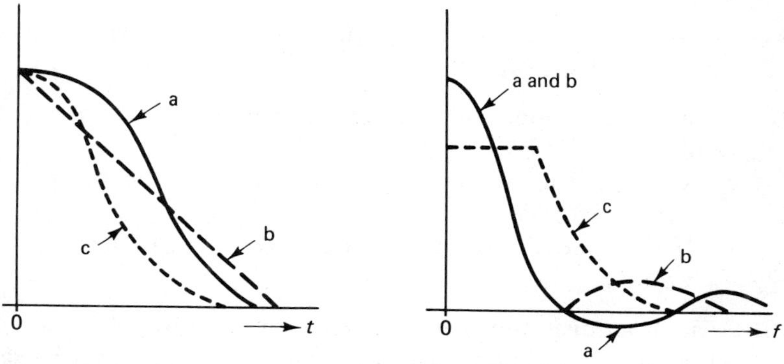

Figure 2-29. Other Windows: (a) Hanning or Raised Cosine, (b) Bartlett, and (c) Parzen

PROBLEMS

2-1. Define the concept of sampling.

2-2. a. Name the three general kinds of signals.
b. What is the basic difference between deterministic and random signals?

2-3. A data acquisition input has a signal-to-noise ratio of 2:1.
a. What theoretical improvement can be realized by averaging over 1,024 samples?
b. What is the effective signal-to-noise ratio?

2-4. Using expressions (2-7) and (2-8) as a basis, develop the expressions of (2-13) and (2-14) for the Fourier integral.

2-5. a. The concept of convolution is concerned with which three mathematical concepts?
b. What three important concepts should you remember when working with the time and frequency domains?

2-6. What mathematical operation defines the act of sampling
a. in the time domain, and
b. in the frequency domain?

2-7. The input to a data acquisition system contains the following discrete frequencies:

52.5 kHz	205 kHz
105 kHz	243 kHz
143 kHz	

If the sample frequency is 100 kHz, what will each input frequency appear as in the output?

2–8. Consider a band-pass signal with a rectangular shape. Prove that $f_S = 2Bw$ at K_{max}.

2–9. A data acquisition system has an aperture time of 10 nsec and a resolution of 12 bits.
  a. What is the highest frequency that can be sampled to within $\pm\frac{1}{2}$LSB?
  b. If the system had an aperture uncertainty of 1 nsec, what would f_{max} be?

2–10. A data acquisition system has an aperture time of 2 nsec. To within what resolution accuracy can a 10-kHz signal be sampled at the point of maximum rate of change?

2–11. A signal is sampled for 120 sec at a 20-kHz sample rate, using a rectangular resolution window.
  a. What is the foldover frequency?
  b. What is the frequency resolution?

REFERENCES

1. K. E. Beauchamp, *Signal Processing*. New York: Halsted Press Book, John Wiley & Sons, Inc., 1973.
2. B. M. Gordon, "Digital Sampling and Recovery of Analog Signals," *EEE Magazine* (May 1970).
3. B. P. Lathi, *Signals, Systems and Communication*. New York: John Wiley & Sons, Inc., 1965.
4. D. W. Mix, *Random Signal Analysis*. Reading, Mass.: Addison-Wesley Publishing Co., Inc., 1969.
5. A. Papoulis, *The Fourier Integral and Its Applications*. New York: McGraw-Hill Book Company, 1962.
6. M. Schwartz, *Information Transmission, Modulation and Noise*. New York: McGraw-Hill Book Company, 1959.
7. M. Schwartz & L. Shaw, *Signal Processing: Discrete Spectral Analysis, Detection and Estimation*. New York: McGraw-Hill Book Company, 1975.
8. W. D. Stanley, *Digital Signal Processing*. Reston, Va.: Reston Publishing Company, Inc., 1975.

3
Quantization and Recovery

3-0 INTRODUCTION

In the previous chapter, we discovered the many complications encountered when performing proper sampling of the various types of analog signals. This chapter completes the sampling picture. It discusses the conversion of the time samples to digital form (quantization) and the subsequent reconstruction either of the original signals or of some derivative (recovery). Like the pieces of an interlocking puzzle, all three functions must fit together properly if the sampled-data system is to produce the desired results. Since, to a great extent, each depends on the other, the proper system solution is a simultaneous optimization of all three. The end of the chapter rounds out the total sampled-data system discussion by providing an example that highlights the trade-offs to be considered.

3-1 QUANTIZATION

The major premise of the data acquisition system is that signals are converted from the analog domain to the digital domain, so that a digital computer or some other digital process may operate on them. This process, called *quantization*, is defined as:

the conversion of an input function that has values in a continuous range to an output that has only discrete values.

Sometimes, quantization is combined with sampling and the two operations occur simultaneously. In other implementations, such as when a sample-and-hold is used, sampling and quantization are two very distinct and separate operations. Here, the sample-and-hold makes the basic measurement and holds the value while the quantizer converts it into a digital format.

The transfer function for a typical quantizer, or analog-to-digital converter, is found in Figure 3-1. While the analog input voltage may be any value between zero volts and the full-scale value, V_{FS}, the output can exist only as one of sixteen discrete values from zero to fifteen. So we find the transfer function to be a staircase. Each of the staircase "flats" and its associated digital output code are referred to as *quantum levels*. The distance between any two quantum levels is called a *quantum interval*. For a binary- (or power-of-2)-based quantizer, the quantum interval, α, is defined as:

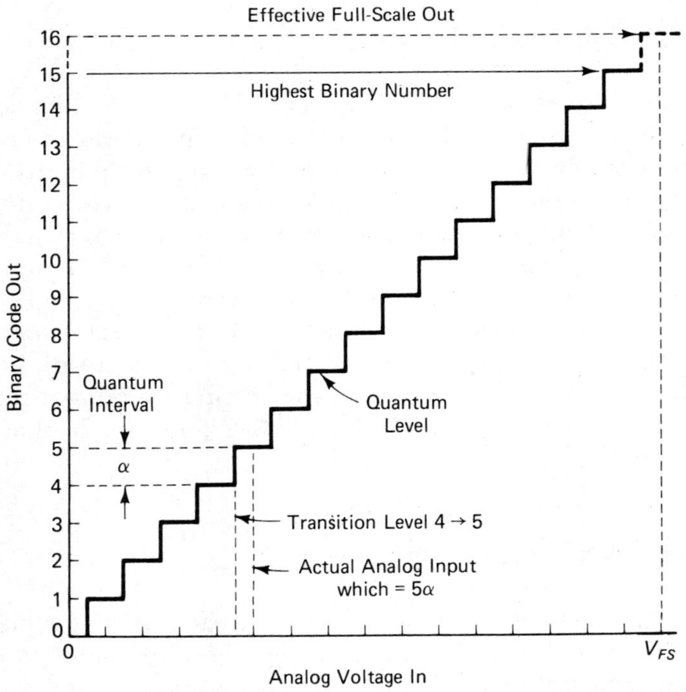

Figure 3-1. Transfer Function for 4-Bit Binary (16-Level) Quantizer

$$\alpha = \frac{V_{FS}}{2^n} \tag{3-1}$$

where V_{FS} = full-scale voltage input
2 = number base for binary
n = number of bits (binary digits)
2^n = number of quantizing intervals

For the quantizer of Figure 3-1, we find that:

$$\alpha = \frac{V_{FS}}{2^n} = \frac{V_{FS}}{2^4} = \frac{V_{FS}}{16}$$

Should 5 bits have been required, α would be:

$$\frac{V_{FS}}{2^5} \quad \text{or} \quad \frac{V_{FS}}{32}$$

and the transfer function staircase would contain twice as many steps as Figure 3-1, and they would be half-sized. The steps continue to double in number and halve in size as the number of quantizer bits increase, as follows:

n	*Steps*
1	2
2	4
3	8
4	16
5	32
6	64
7	128
8	256
9	512
10	1,024
11	2,048
12	4,096
13	8,192
14	16,384
15	32,768
16	65,536

Referring to Figure 3-1, note that level 16 does not actually exist and that the highest binary number output available is level 15. It should be mentioned that the voltage range of 0 volts to V_{FS} was selected for simplicity. In actuality the quantizer input could range from any arbitrary low value to any arbitrary high value.

Several binary codes are commonly used. The most common are the straight binary and offset binary, which are found in Table 3-1. *Straight binary* is a unipolar code where 0000 is equivalent to level 0 (also analog zero), while 1111 is equivalent to level 15. On the other hand, *offset binary* is a bipolar code with 0000 equal to −7, and 1111 equal to +7. For this situation, the 2^3 bit (most significant) is the sign bit, with 0 = − and 1 = +. There are two "zero" levels in the offset binary code, each half a quantum interval from true analog zero.

Two other codes are more compatible with standard digital computers (Table 3-2). They are the one's and two's complement binary codes. Both bipolar, they differ primarily in the positioning of the true analog zeros. The one's complement code is identical with the offset binary code with the exception that the sign bit is inverted. In this case, 1 is *negative* and 0 is *positive*. Like offset binary, this system has

Table 3-1 Straight and Offset Binary Codes

	Bit Position				*Level*	
	2^3	2^2	2^1	2^0	*Straight Binary*	*Offset Binary*
Straight Binary Zero →	0	0	0	0	0	−7
	0	0	0	1	1	−6
	0	0	1	0	2	−5
	0	0	1	1	3	−4
	0	1	0	0	4	−3
	0	1	0	1	5	−2
	0	1	1	0	6	−1
	0	1	1	1	7	−0
Offset Binary Zero →						
	1	0	0	0	8	+0
	1	0	0	1	9	+1
	1	0	1	0	10	+2
	1	0	1	1	11	+3
	1	1	0	0	12	+4
	1	1	0	1	13	+5
	1	1	1	0	14	+6
	1	1	1	1	15	+7

Table 3-2 One's and Two's Complement Binary Codes

	Bit Position				*Level*	
	2^3	2^2	2^1	2^0	*Two's*	*One's*
	1	0	0	0	−8	−7
	1	0	0	1	−7	−6
	1	0	1	0	−6	−5
	1	0	1	1	−5	−4
	1	1	0	0	−4	−3
	1	1	0	1	−3	−2
	1	1	1	0	−2	−1
	1	1	1	1	−1	−0
One's Zero →						
Two's Zero →	0	0	0	0	0	+0
	0	0	0	1	+1	+1
	0	0	1	0	+2	+2
	0	0	1	1	+3	+3
	0	1	0	0	+4	+4
	0	1	0	1	+5	+5
	0	1	1	0	+6	+6
	0	1	1	1	+7	+7

two zero levels. The two's complement code, however, has only one zero level, and it is coincident with the 0000 code. There is also one more negative level (−8) than positive (+7). The sign is also the complement of that for offset binary. Careful examination shows that all three bipolar codes discussed thus far are simple extrapolations from straight binary requiring either offsetting of analog zero with respect to the code levels and/or simple inversion of the sign bit.

Another binary-type code, which was popular in the late seventies, is the absolute-value-plus-sign code shown in Table 3-3. The unique feature of this code is that the lower three bits are exactly the binary representation of the associated level; for example, 111 = 7. As a result, the lower three bit codes reflect about the zero level; that is, +3 has the same code as −3. Like the other bipolar codes, the most significant (2^3) bit is the sign designator.

A code used by some quantizers is the Gray code. This code, shown in Table 3-4, has two unique features. First, like the absolute-value-plus-sign code, the lower three bits reflect about the scale midpoint. Second, as the code progresses from one level to the next, only one bit changes. The Gray code is used for most types of electromechanical quantizers and for cyclic-type A/D converters.

Table 3-3 Absolute-Value-Plus-Sign Code

	Bit Position				*Level*
	2^3	2^2	2^1	2^0	
	0	1	1	1	−7
	0	1	1	0	−6
	0	1	0	1	−5
	0	1	0	0	−4
	0	0	1	1	−3
	0	0	1	0	−2
	0	0	0	1	−1
	0	0	0	0	−0
Zero ⟶					
	1	0	0	0	+0
	1	0	0	1	+1
	1	0	1	0	+2
	1	0	1	1	+3
	1	1	0	0	+4
	1	1	0	1	+5
	1	1	1	0	+6
	1	1	1	1	+7

One other commonly used code is the 8421 binary-coded decimal (BCD), as shown in Table 3-5. This code is utilized when direct decimal readouts are required. Although the codes are binary in form, they exist only in decades (0 to 9). When the limits of a decade are exceeded, the code carries into the next higher decade. Each successive decade is identical to that shown in the table, but it has a magnifier of 10 associated with it. The quantum interval for the 8421 BCD code is calculated as follows:

$$\alpha = \frac{V_{FS}}{10^m} \tag{3-2}$$

where V_{FS} = full-scale voltage input
2 = number base for decimal
m = number of decades (or digits)
10^m = number of quantizing intervals

Now let's get back to the quantization process itself.

An expansion of a segment of Figure 3-1 is found in Figure 3-2. Here we see the transitions between the given levels $K - 1$ and K and between levels K and $K + 1$. Of interest here is the fact that, once the

Table 3-4 The Gray Code

	Bit Position				*Level*
	2^3	2^2	2^1	2^0	
	0	0	0	0	0
	0	0	0	1	1
	0	0	1	1	2
	0	0	1	0	3
	0	1	1	0	4
	0	1	1	1	5
	0	1	0	1	6
	0	1	0	0	7
Midpoint →					
	1	1	0	0	8
	1	1	0	1	9
	1	1	1	1	10
	1	1	1	0	11
	1	0	1	0	12
	1	0	1	1	13
	1	0	0	1	14
	1	0	0	0	15

output code has made the transition from level $K - 1$ to level K, it does not change again until the analog input has progressed from $V_K - \alpha/2$ to $V_K + \alpha/2$, or a value equivalent to a full quantum interval. So obviously, when the quantizer output level is K, all we can say about the input is that it is somewhere between $V_K - \alpha/2$ and $V_K + \alpha/2$. This element of uncertainty produces an irreducible error situation. The only time the output code is totally accurate is when the

Table 3-5 The 8421 BCD Code

8	*4*	*2*	*1*	*Level*
0	0	0	0	0
0	0	0	1	1
0	0	1	0	2
0	0	1	1	3
0	1	0	0	4
0	1	0	1	5
0	1	1	0	6
0	1	1	1	7
1	0	0	0	8
1	0	0	1	9

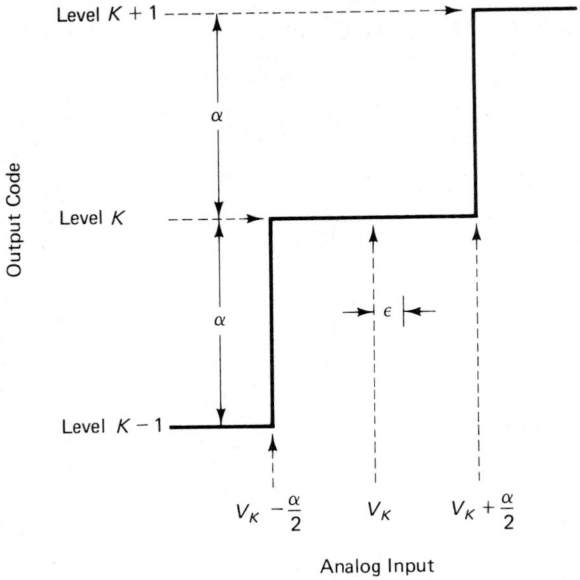

Figure 3-2. Uncertainty at Quantizer Output

input is exactly V_K. Having the output code remain constant, while the input ranges over all possible values $V_K \pm \alpha/2$, is the same as adding noise to the quantizer input. This effect is called *quantization error* or *quantization noise*. If we consider any given analog input, $f(t)$, to occur over a long period, it is probable that all values of $V_K \pm \alpha/2$ will be present and will cause the Kth level to occur in a random fashion. The amplitude of the effective noise is established by the number of quantizer bits or quantum intervals. If the input deviation from the voltage value equivalent to the Kth level, V_K, is designated as ϵ (Figure 3-2), the RMS quantization error is calculated as follows:

$$\sqrt{\frac{1}{\alpha}\int_{-\alpha/2}^{\alpha/2} \epsilon^2 d\epsilon} = \sqrt{\frac{\alpha^2}{12}} = \frac{\alpha}{2\sqrt{3}} \qquad (3\text{–}3)$$

The character of the quantization error function is found in Figure 3-3. As you can see, the error is triangular in nature, and it cycles back and forth between $+\alpha/2$ and $-\alpha/2$ as the analog input proceeds up or down the scale. Quantization error has a probability function that is rectangular in nature as, over a large period, the input is as likely to occur at any one value of ϵ as any other, while the output remains fixed. This situation is true for the perfect quantizer. Yet, as Gordon reminds us (reference 5), quantizers add noise themselves from a number of internal sources. This noise, generally Gaussian in character, has a definite impact on the effective quantization error rate.

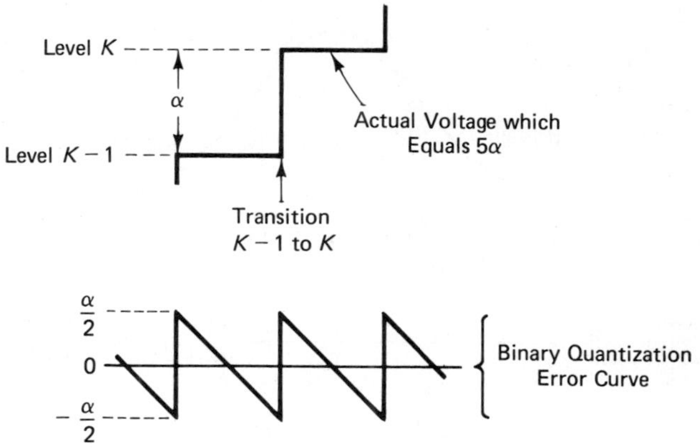

Figure 3-3. Expanded Section of Binary Quantizer Transfer Function with Error Curve

Figure 3-4 depicts this situation when the analog input is centered in the quantum interval. This figure shows that the noise peaks above $\pm 2.5\sigma$ extend into the adjacent quantum intervals and that 1.25 percent of the time the output code is in error. Unfortunately, as discussed, the input does not always occur at the center of the interval, and it is just as likely to be at any other value.

Figure 3-5 gives an indication of what happens when the input is offset from the center. Although the error output favors the $K + 1$ side, it has also substantially increased in value. The quantizer output will be in error 6.7 percent of the time under these conditions. This type of result occurs when the RMS quantizer noise is only 20 percent of a quantum interval. You should therefore consider the noise figure of the analog-to-digital converter (quantizer) when making a selection.

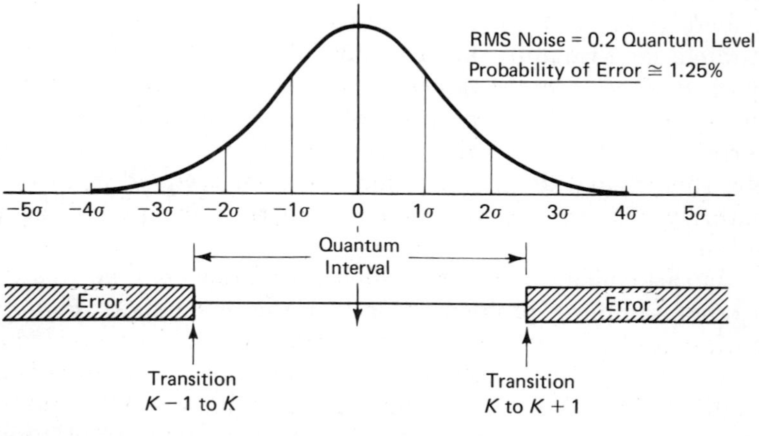

Figure 3-4. Noise Centered in the Quantum Interval

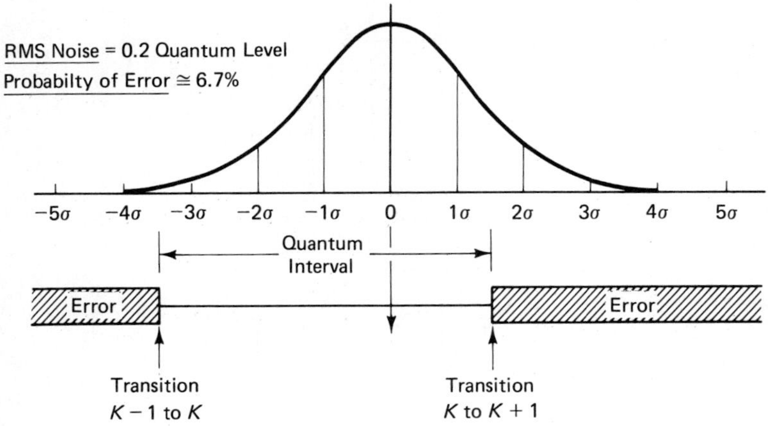

Figure 3-5. Noise Offset in the Quantum Interval

3-2 QUANTIZATION AND RANDOM SIGNALS

In Chapter 2, we recognized that a certain class of dynamic signals is characterized by its statistics. These signals are defined by their probability density functions rather than by their absolute values at particular points in time. As it turns out, the requirements for quantizing resolution for this type of signal is much less stringent than it is for signals that we must precisely define at every point.

Consider the relationship shown in Figure 3-6. Figure 3-6a is the probability density function for the time/voltage signal of Figure 3-6b. The horizontal lines in 3-6b indicate the centers of the quantization intervals. As we see, the centers of the quantization intervals have been extended leftwards to the density function. That this representation is accurate is confirmed by the fact that the quantizer outputs can occur only at specific discrete values. The quantized density function is, then, a measure of the number of times that each discrete value occurred with respect to the total number of samples taken. Effectively, the original density function has been *sampled* in the voltage domain by a sample function that occurs at a rate equal to the quantum interval.

As this conclusion implies, such a situation may be treated like any other sampling problem. Consider the elements of Figure 3-7. Translation of the probability density function in the voltage domain to its complementary "spectrum" in the ν-frequency domain is found in Figure 3-7a. Figure 3-7b shows the translation of the quantizing function from one domain to the other. Because these translations are

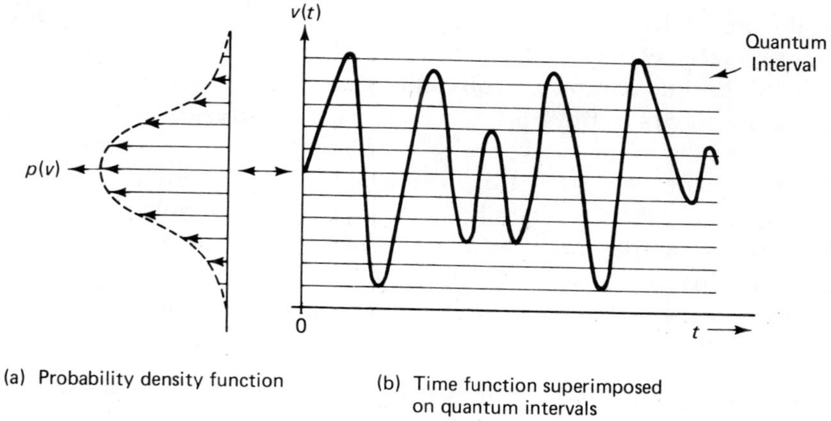

Figure 3-6. Relationship of Quantization and Probability

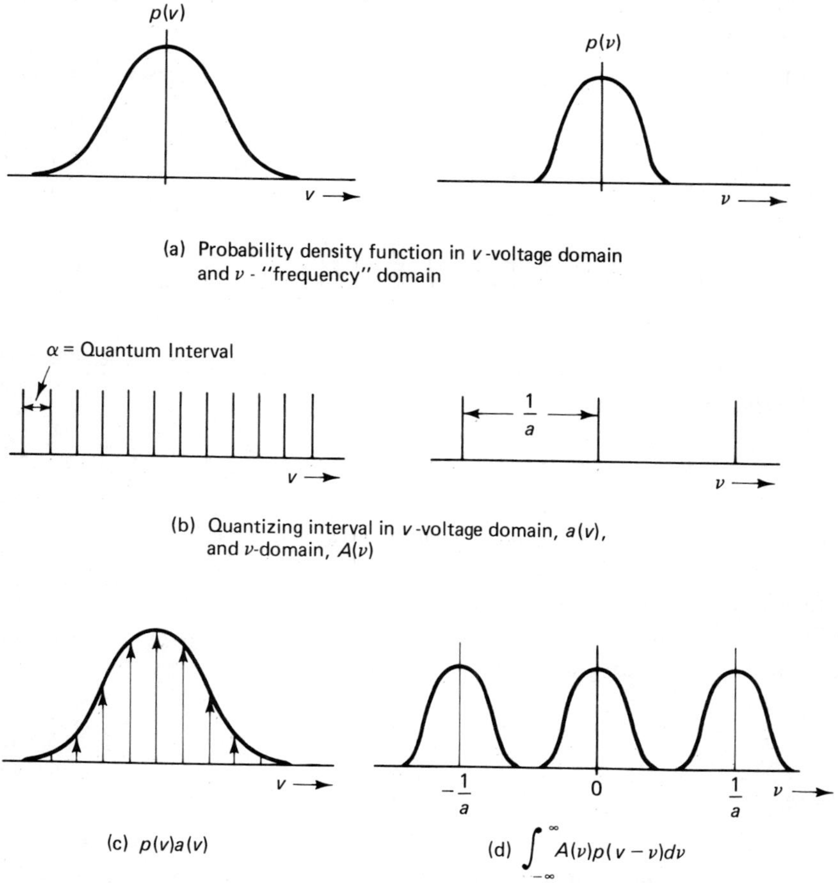

Figure 3-7. Quantization and Probability in Two Domains

analogous to the sampling of a time signal with an impulse sample function (see Section 2–2), the results of Figure 3–7c and 3–7b are not surprising. The sampled probability density function in the voltage domain appears as multiple "spectra" in the ν domain. All the other sampling rules hold as well, and, if we have fewer quantization levels (a increases), the "spectra" of 3–7d move closer together. If they should overlap, you would not be able to distinguish one level from another because the sample rate would be too low. So the principles of aliasing apply in this case as well. As you would expect, there is a quantizing theorem analogous to the sampling theorem (two samples per cycle ...). The rule of thumb goes:

> If a quantizing width is chosen such that the range of $V(t)$ is at least eight quantum intervals, then the first few statistical moments of the quantized signal may be used as moments of $V(t)$, providing Sheppard's corrections are used (reference 1).

To see how this applies, consider the first two moments as shown:

$$m_1 = \text{Average value} = \int_{-\infty}^{\infty} Vp(V)dV \qquad (3\text{–}3a)$$

$$m_2 = \text{Second moment} = \int_{-\infty}^{\infty} V^2p(V)dV \qquad (3\text{–}3b)$$

The mean-squared and root-mean-squared (RMS) values of V are further defined as:

$$\sigma^2 = m_2 - m_1^2 \qquad (3\text{–}5)$$

and:

$$\sigma = \sqrt{m_2 - m_1^2} \qquad (3\text{–}6)$$

Often very important are the statistical quantities, m_1, σ, and σ^2, when defining a random or nondeterministic signal. Recovery of these parameters by use of Sheppard's corrections is:

$$\text{Average value} = \mu_1 = \int Vp(V)^*dV \qquad (3\text{–}7)$$

where $p(V)^*$ = the quantized and sampled result

$$\text{Second moment} = \mu_2 - \frac{1}{12}\alpha^2 \qquad (3\text{–}8)$$

$$\text{Mean-square value} = \mu_2 - \frac{1}{12}\alpha^2 - \mu_1^2 \qquad (3\text{-}9)$$

$$\text{RMS value} = \sqrt{\mu_2 - \frac{1}{12}\alpha^2 - \mu_1^2} \qquad (3\text{-}10)$$

where μ_1 and μ_2 = the simple moments calculated from the quantized samples

α = the quantum interval

$\frac{1}{12}\alpha^2$ = the correction

Two observations can be made. First, the first moment or average value is the same whether quantized or not. Second, with a small increase in the number of quantum levels, no corrections are needed. One final comment on statistical quantization. Due to a corollary to sampling, the frequency content of the probability density function is important. Specifically, signals with favorite points or values that occur frequently must be handled carefully. This type of signal—such as single sinusoid—has a high "frequency" content, and the quantizing rule of thumb does not apply.

3-3 SIGNAL RECOVERY

In the first part of Chapter 2, it was implied that the sample data process was not complete until a function called *recovery* was effected. It was further suggested that this recovery function was necessary to reconstruct either the original time signal, or a derivative of it, after digital processing. Why is a recovery or reconstruction process necessary? There are two obvious reasons:

1. errors of omission, and
2. errors of commission.

For an error of omission, look at Figure 3-8. The dashed line represents the envelope of the original time function. The result of sampling, however, is a series of values equal to the sample points depicted by the arrow heads. At this point, a great deal of the signal trend information is missing. Once quantization has been added and the samples can have only discrete values, the difficulties are compounded.

The second reason for recovery—errors of commission—is found in Figure 2-12c of Chapter 2. After sampling, not only is the original

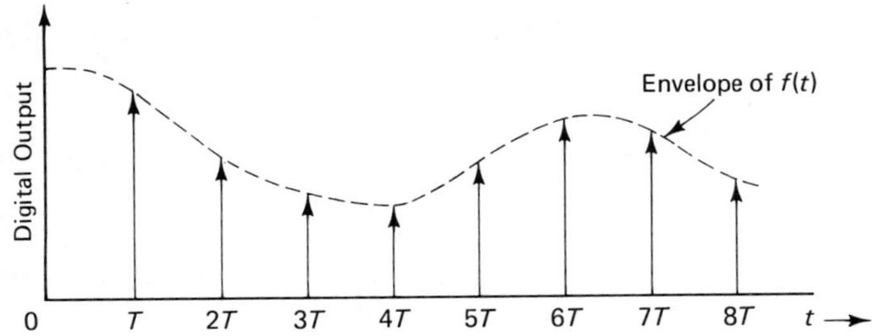

Figure 3-8. Typical Digital Processor Output, $f^*(t)$

signal spectrum present, but any number of additional spectra are also present and centered about multiples of the sample frequency. If these are not eliminated, they may cause totally erroneous results within the digital process. They will surely prove confusing in a digital communications system.

A number of different recovery operators are used, and we will discuss some of the more common ones. But first, let's consider the ideal recovery function, shown in Figure 3-9. The function is a rectangular or "box" filter in the frequency domain. When this function is translated to the time domain, it takes the form $(\sin x)/x$. Also, you should not be too surprised to find that recovery is the inverse function to sampling. Whereas sampling is multiplication in the time domain and therefore equivalent to convolution in the frequency domain, recovery is just the opposite. Utilizing the ideal recovery operator, then, is equivalent to convolving the $(\sin x)/x$ function of Figure 3-9 with the time samples of Figure 3-8.

The result is shown in Figure 3-10. If the parts of the various $(\sin x)/x$ functions are added together, the original waveform would be reconstructed. How closely the original function is reproduced would depend on the granularity of quantization. In Chapter 2, it was mentioned that the validity of the sampling theorem depended on the use of the ideal rectangular operator. To validate this statement, consider

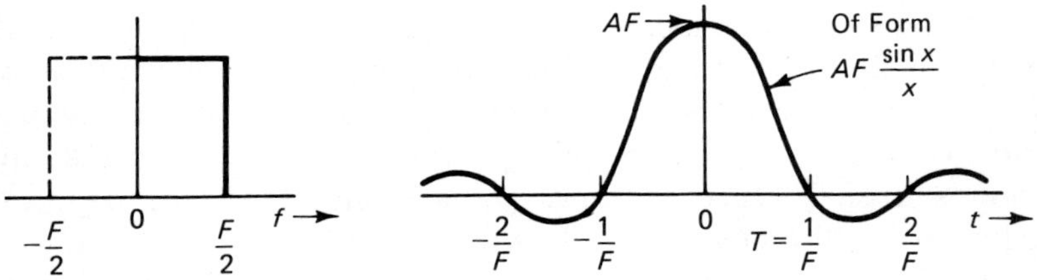

Figure 3-9. Ideal Recovery Function, Where $T = 1/F$

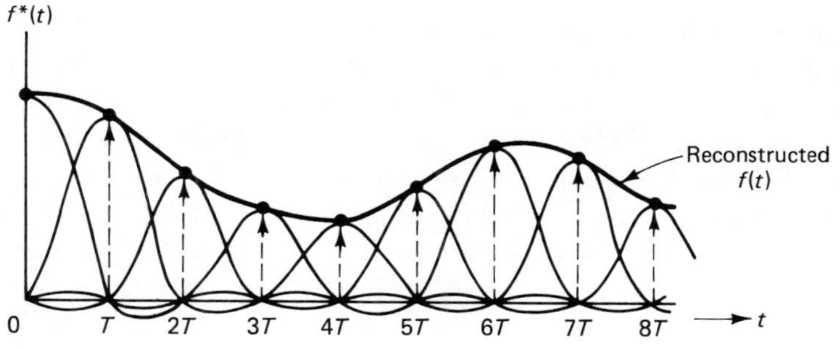

Figure 3-10. Ideal Signal Recovery

the two sample points from a single sinusoid. Now convolve the sample with the (sin x)/x function. The result is a reproduced sine wave.

Unfortunately, it is not physically possible to implement a filter with the ideal recovery function characteristics. In fact, the most commonly used reconstructing function has exactly the opposite characteristics. This is the zero-order hold function shown in Figure 3-11. As seen in Figure 3-11a, the time and frequency domain representations for the zero-order hold are the exact inverses of those for the ideal operator. The function is rectangular in the time domain and (sin x)/x in the frequency domain. Figure 3-11b provides the final convolved

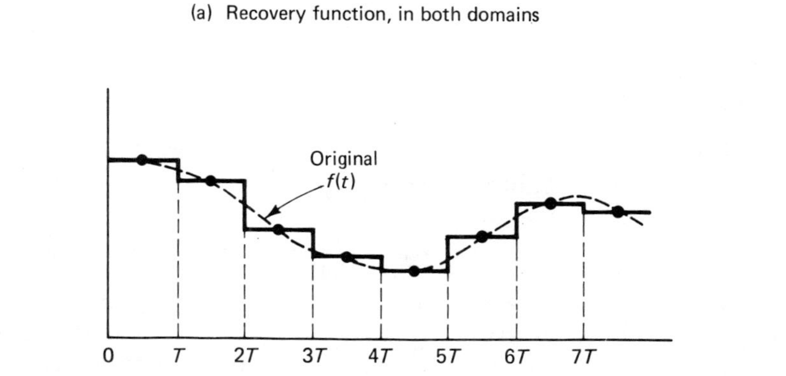

(a) Recovery function, in both domains

(b) Reconstructed function, $f(t)$

Figure 3-11. Staircase or Zero-Order Hold Recovery

result in the time domain. This output is quickly recognized as that of a typical digital-to-analog (D/A) converter. Unlike the results shown in Figure 3-10, this reconstruction process results in substantial error. The worst case for this error would occur at the zero crossing of the sinusoid for the highest frequency contained in the original signal.

This error is diagrammed in Figure 3-12. Let's analyze this error. The maximum slope of the sinusoid, $V(t) = A \sin \omega t$, occurs at $t = 0$ and is:

$$\text{Maximum slope} = \left.\frac{dV(t)}{dt}\right|_{t=0} = A\omega$$

Using the terms of Figure 3-12, we also find that

$$\text{Slope} = A\omega = \frac{V_{\text{error}}}{T}$$

where T = the sample period, and therefore

$$V_{\text{error}} = A\omega T$$

The percentage of error is:

$$\%\ \text{error} = \frac{V_{\text{error}}}{A} \times 100 = \frac{A\omega T}{A} \times 100\ \omega T$$

or, as $\omega = 2\pi f$:

$$\%\ \text{error} = 200\ \pi f T \qquad (3\text{–}11)$$

Recognizing that $1/fT$ = samples/cycle, we can develop the expression:

$$\text{Samples/cycle} = \frac{200\pi}{\%\ \text{error}} \qquad (3\text{–}12)$$

Utilizing this expression, we find that you have to sample the original waveform by at least 628 samples/cycle to achieve an instantaneous reconstruction error of less than 1 percent. So now we see that the sample rate is directly affected by the type of reconstruction operator selected.

Another way to look at this situation is to consider that we have applied a narrow-band (sin x)/x filter to the original frequency spec-

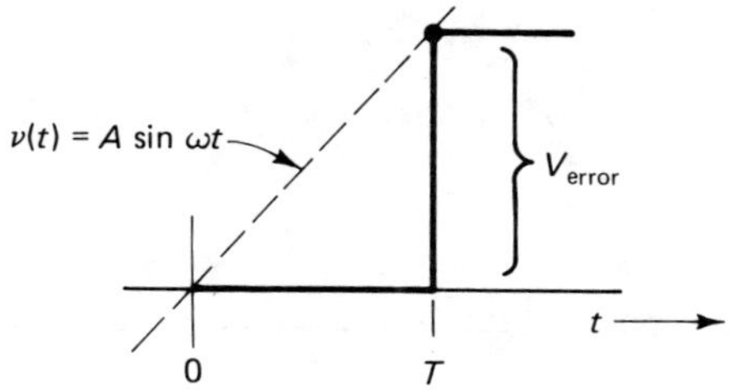

Figure 3–12. Zero-Order Hold Error Function

trum. One method of solving this problem with feasible hardware is to provide a narrow return-to-zero output from the D/A converter and to follow it with sharp cutoff analog filter. This combination, shown in Figure 3–13, moves the first zero crossing, $1/\tau$, out in frequency so that the original input spectrum is minimally affected by the $(\sin x)/x$ filtering action. The analog filter provides the storage to fill in the holes between the sample points.

In many situations, a digital computer is used to fill in or to interpolate between the sample points, as illustrated in Figure 3–14. This method of recovery is called the *predictive first-order hold* and the expression for the recovery operator is:

$$X(t) = \frac{[X(n) - X(n-1)]}{T}[t - nT] + X(n) \tag{3–13}$$

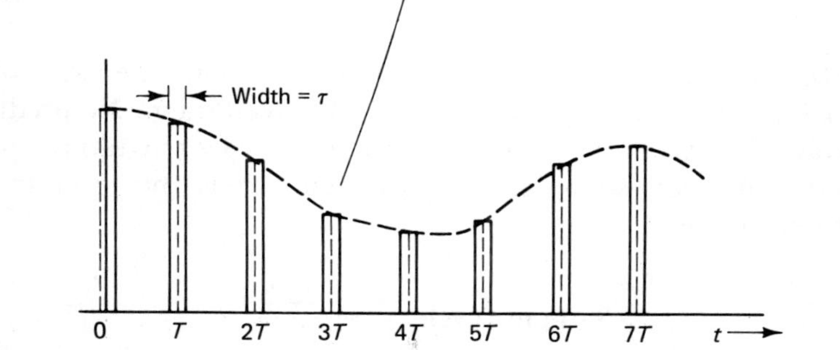

Figure 3–13. Recovery Using D/A Converter with RZ Output Followed by an Analog Filter

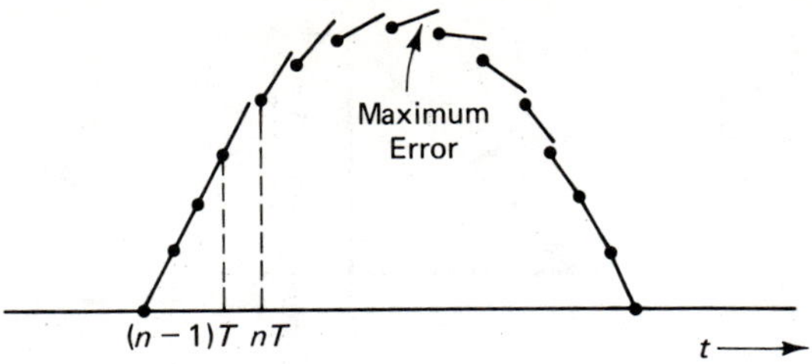

Figure 3-14. Predictive First-Order Hold

At any given sample point, the computer calculates the slope between the previous sample and the present sample and assumes the slope between the next two points will be the same. This works well as long as the slope is not changing. When it is changing, such as at the top of the sinusoid, substantial errors start to creep in. In terms of the instantaneous error percentage, the samples per cycle for this recovery operator are:

$$\text{Samples/cycle} = \frac{62.8}{\sqrt{\%\ \text{error}}} \tag{3-14}$$

The results are substantially better than that for the zero-order hold. Yet you must exercise care when using a predictive reconstructor in any closed-loop system. An erroneous assumption can cause the loop to become unstable.

Another first-order recovery operator, the linear first-order hold, can be seen in Figure 3-15. The expression for this recovery function is:

$$X(t) = \frac{[X(n) - X(n - 1)]}{T}[t - nT] + X(n - 1) \tag{3-15}$$

In this case, the computer waits until it knows the value of any two points and then simply draws a line between them. No prediction is involved in this process, but the output is always one sample period behind. The expression for samples/cycle versus the instantaneous error percentage is:

$$\text{Samples/cycle} = \frac{22.3}{\sqrt{\%\ \text{error}}} \tag{3-16}$$

Often, to smooth out the recovered output, the recovery algo-

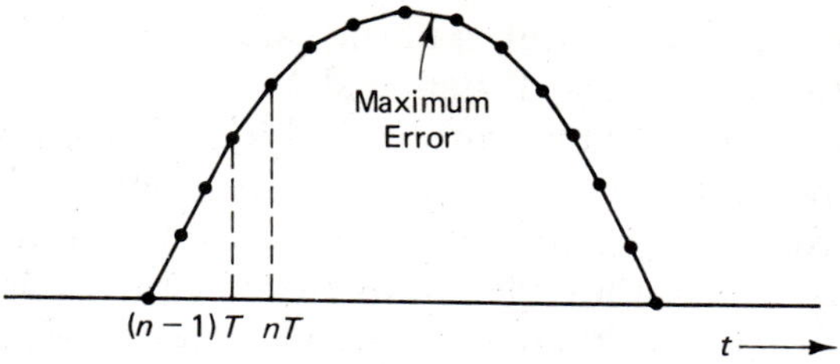

Figure 3-15. Linear First-Order Hold

rithm inserts intermediate points between the actual samples. For example, where accurate recovery is very critical, second-order functions are sometimes used. This procedure uses not only the differences between the last two samples, but also the difference between that difference and the difference between the previous two samples. Second-order computations may be effected for both the linear and predictive functions.

Sample rates for the worst-case, instantaneous error result from expressions (3-12), (3-14), and (3-16). In many practical cases, the root-mean-square (RMS) error is more useful. The RMS error is the instantaneous error averaged over time. Table 3-6 provides a comparison of sample rates versus three orders of RMS recovery error. The recovery operators compared are zero-order hold (step), first-order linear, and 2-, 3-, and 4-pole Butterworth filters. This particular table is based on the input data having the characteristics of a fourth-order

Table 3-6 Recovery Error Versus f_S(Sample Frequency)/$f_{3\ db}$ of Input Spectrum

	Percentage of RMS Recovery Error		
Recovery Operator	*0.01*	*0.1*	*1.0*
Zero-order (step)	12,000	1,200	120
Linear	160	48	16
Butterworth ($n = 2$)	1,920	240	30
Butterworth ($n = 3$)	280	72	18
Butterworth ($n = 4$)	108	36	12

Note: Based on input having the characteristics of a fourth-order Butterworth spectrum.

(4-pole) Butterworth spectrum, and the values on the chart have units in samples per cycle of a 3-dB bandwidth.

A similar chart could be constructed for other shapes and orders of possible input data spectra. This point highlights one of the great difficulties involved with data acquisition. There are nearly as many shapes and orders of input spectra as there are data acquisition systems. So only a few "cookbook" charts, like Table 3-6, are available, and they cover only a particular set of circumstances.

3-4 A DATA ACQUISITION EXAMPLE

At this point an example is appropriate, to tie the main concepts of Chapters 2 and 3 together. To do so, consider the digital communications system of Figure 3-16. The analog input has the characteristics of a fourth-order Butterworth spectrum, and for this exercise we assume that the highest frequency of interest is at 20 kHz. Requirements for the system are:

1. dynamic range >70 dB
2. aliasing error <0.1 percent
3. recovery error <0.1 percent RMS
4. aperture error <0.1 percent average

Things to be determined are:

1. number of quantization levels (bits),
2. sample rate,
3. sampling aperture, and
4. recovery operator.

Let's first consider the number of quantization levels or bits required. The system dynamic range has been specified as 70 dB. This is equivalent to the ratio 3,162:1. The nearest greater binary number is 4,096, or 12 bits, which is equivalent to 72 dB. (Note: you might consider a 13-bit quantizer, which would provide a 78-dB dynamic range.)

Selection of the sampling rate is a bit more complicated because we have two requirements: one for aliasing and one for recovery error. Let's consider the aliasing problem first. Simply stated, it requires the error due to folding to be less than 0.1 percent of the expected value at the highest frequency of interest, 20 kHz. As we see in Figure 3-16, at 20 kHz, the signal levels are 4.4 percent of the dc value. The overlap of the first harmonic spectrum must not exceed 0.1 percent of the 4.4 percent value. This overlap occurs when the first harmonic spectrum

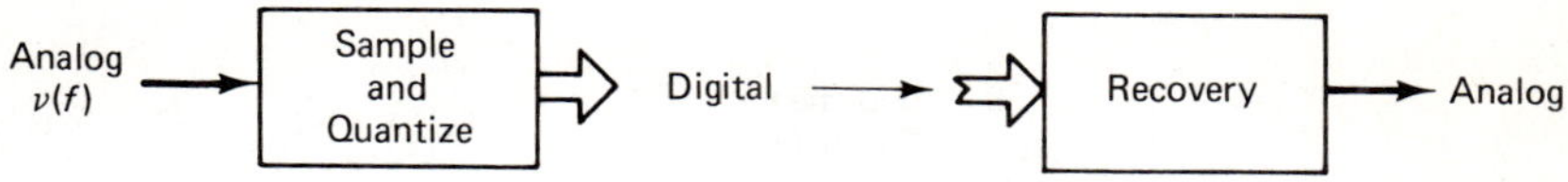

(a) Example sample data system

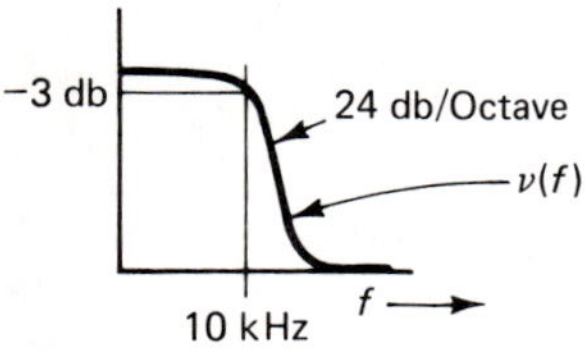

$$\nu(f) = \frac{A}{\sqrt{1 + \left[\dfrac{f}{f_{3\,db}}\right]^8}}$$

f	$\%A$	Attenuation
10 kHz	70.7	−3 dB
20 kHz	4.4	−27 dB
40 kHz	0.28	−51 dB
80 kHz	0.018	−75 dB
120 kHz	0.0044	−87 dB
160 kHz	0.0011	−99 dB

(b) Characteristics of fourth order input

Figure 3–16. Data Acquisition Example

contribution of f_S − 120 kHz coincides with the 20-kHz point on the original spectrum. This situation is shown in Figure 3–17. As expected, there is a 1,000:1, or 60-dB, difference between the two spectra. The sample rate to satisfy this requirement must then be equal to or greater than:

$$120 \text{ kHz} + 20 \text{ kHz} = 140 \text{ kHz}$$

Before we firmly establish 140 kHz as the sample rate, consider the impact of the recovery operator. To make things simple, let's choose the recovery filter to have the same characteristics as the input spectrum, namely a 4-pole Butterworth. Fortunately, Table 3–6 includes this type of filter for 0.1 percent recovery error. We quickly see that:

$$\frac{f_S}{f_{3\text{ dB}}} = 36$$

and, for our situation:

$$f_S = 36 \times 10 \text{ kHz} = 360 \text{ kHz}$$

So, for this particular case, the recovery requirements dictate the sample rate. At f_S = 360 kHz, aliasing is negligible. A good selection for this situation would be a 12-bit, 500-kHz, successive approximation

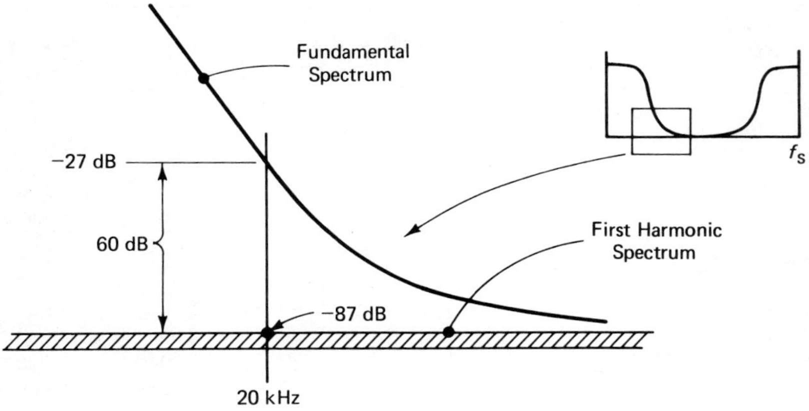

Figure 3–17. Effect of Aliasing, $f_S = 130$ kHz

A/D converter. With this choice established, the next decision concerns whether to use a sample-and-hold function, prior to the converter. The basic aperture of a 500-kHz A/D converter is:

$$\frac{1}{500\ \text{kHz}} = 2\ \mu\text{sec}$$

Such an aperture would be equivalent to placing a $(\sin x)/x$ filter in the system, with the first zero crossing at 500 kHz. This insertion would cause between 1- and 2-percent attenuation at 20 kHz. To obtain a degradation of less than 0.1 percent, a sample-and-hold with an aperture less than 100 nsec would be necessary.

Note that we have been satisfied with RMS values of accuracy. Had a case been made for instantaneous accuracy at any point on the highest frequency of interest sinusoid, the aperture expressions of Section 2–6 would apply, and the required aperture uncertainty would be a few nanoseconds.

After the digital data has been transmitted from the acquisition point to the point of reception, recovery must be considered. We have already chosen the final recovery filter to be a 4-pole Butterworth, but first the data must be converted from digital binary numbers to analog values. Suppose we utilized a simple digital-to-analog converter (zero-order hold), with an update rate equal to the original conversion rate of 500 kHz, prior to the filter. We still have the problem of $(\sin x)/x$ filtering, which in this case would be more severe than the final filter. A way to solve the problem would be to utilize a D/A converter with return-to-zero outputs (Figure 3–13) of less than 100 nsec in width.

The final system configuration is shown in Figure 3–18. An

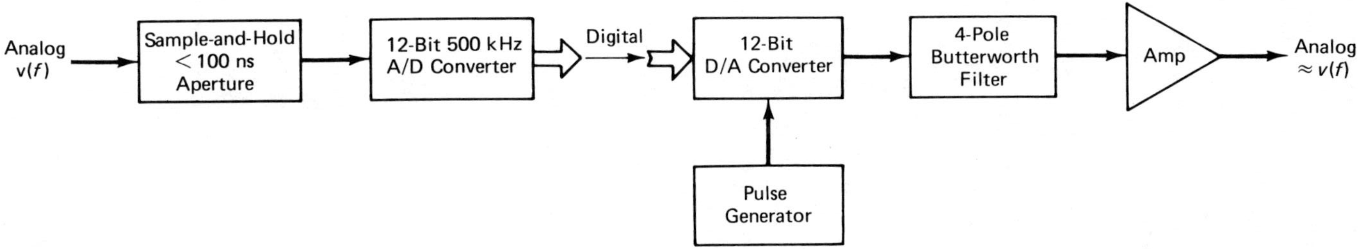

Figure 3–18. Complete Data Acquisition System

amplifier is utilized in the output to compensate for the loss incurred when low-energy pulses are applied to the complex filter.

3-5 SUMMARY OF DESIGN CONSIDERATIONS

There are, of course, many potential solutions for the example of Section 3-4. An arbitrary decision was made to utilize the 4-pole Butterworth recovery filter. Utilizing a more complex filter or adding interpolation points by computer could have substantially reduced the required sample rate. Increasing the aliasing error requirements, on the other hand, could have pushed the sample rate in the other direction. Each specific problem must be considered on its own merits, based on the available or feasible hardware. There are no hard-and-fast, *safe* rules of thumb. Designers are also fortunate if they can find charts, tables, or curves to let them "cookbook" their problems. Unfortunately, too many diverse problems need solving to allow for such design aids. Beware of the charts and curves that do exist, since their authors are most often constrained to select a perfect, classical situation; so such charts seldom have widespread usage. To evolve a successful design, you must know what the input signal spectrum looks like, what you wish to obtain from it, and how to apply the rules of sampling, quantization, and recovery on an individual basis.

PROBLEMS

3-1. A quantizer has a full-scale input range of +10 V and a resolution of 13 bits.
  a. What is the value of a quantum interval?
  b. Suppose the quantizer had a resolution of 4 BCD digits. What is the value of a quantum interval?

3-2. Consider Figure 3-1. Why do the transition points from one output code to the next occur at the midpoints of the quantization intervals?

3-3. Using the results of problem 3-1, and considering no internal quantizer noise, what is the RMS quantization noise?

3-4. A simple sinusoid has an average value, a mean-square value, and an RMS value.
Does it lend itself to the quantization "rule of thumb" in Section 3-2? Why?

3–5. The highest frequency in a signal being reconstructed is 20 kHz. The recovery requirements dictate that the maximum instantaneous error be less than 0.1 percent. What minimum sample rates would be required using:
a. the zero-order hold,
b. the first-order predictive hold, and
c. the first-order linear hold?

3–6. The 3-dB point of a fourth-order Butterworth spectrum being recovered is 20 kHz. If an RMS recovery error of 0.1 percent is allowable, what would the minimum sample rates be if you used:
a. the zero-order and
b. the linear recovery operator?
c. Compare the results of a and b with those of problem 3-5.

3–7. Develop a system solution for the data acquisition example of Section 3-4. But utilize the following systems requirements:
a. dynamic range >58 dB
b. aliasing error <0.2 percent
c. recovery error <1-percent RMS
d. aperture error <0.5 percent instantaneous at 20 kHz

REFERENCES

1. Alfred K. Susskind, ed., *Notes on Analog–Digital Conversion Techniques*. New York: Technology Press, John Wiley & Sons, Inc., 1960.
2. D. H. Shiengold, *Analog–Digital Conversion Handbook*. Norwood, Mass.: Analog Devices, 1972.
3. B. M. Gordon, "Digital Sampling and Recovery of Analog Signals," *EEE Magazine* (May 1970).
4. John V. Wait, "Sampled Data Reconstruction Errors," *Instruments and Control Systems* (June 1970).
5. B. M. Gordon, "Noise Effects on Analog to Digital Conversion Accuracy," *Computer Design* (March/April 1974).
6. Herman Schmid, *Electronic Analog/Digital Converters*. New York: Van Nostrand Reinhold Company, 1970.

4 System Parameters

4-0 INTRODUCTION

To properly design or specify a data acquisition system, you must thoroughly understand system parameters. Such knowledge is particularly important because overspecification can prove costly and underspecification is certain to provide less-than-satisfactory operational results.

While many early data acquisition equipments were poorly specified, such is not the case today. So many excellent data sheets exist that the uninitiated could easily be overwhelmed by their contents. Once they understand the basic parameters, the engineer or user need only establish the conditions under which they were determined or measured by the manufacturer.

The purposes of this chapter are to examine the major system parameters and to provide you with a clear understanding of each.

4-1 ABSOLUTE ACCURACY

Absolute accuracy is an indication of the discrepancy between the data acquisition system measured value of a given input and the National Bureau of Standards measured volt. This implies that the system's voltage reference is traceable to the NBS standards. Absolute accuracy is generally not guaranteed unless the system is periodically calibrated.

4-2 RELATIVE ACCURACY

Relative accuracy is a measure of the data acquisition system's ability to yield correct output codes, for all possible input voltages, relative to its full-scale range. It is a direct function of the system's linearity. A system could very easily have excellent relative accuracy characteristics while exhibiting poor absolute accuracy. Such a situation would exist if the system transfer function had been shifted by a fixed input bias. In this case, the measured inputs would produce output codes that are correct only relative to one another. Each output code would, however, be in error by the amount of the added bias, with respect to the input being measured.

The most typical way of stating accuracy is:

$$\underbrace{\pm 0.01\% \text{ of FSR}}_{\substack{\text{Analog} \\ \text{Inaccuracies}}} \qquad \underbrace{\pm \tfrac{1}{2}\text{LSB}}_{\substack{\text{Quantization} \\ \text{Error}}}$$

where FSR = full-scale reading (often shortened to "FS")
LSB = least significant bit

The first term includes all the analog sources of error such as linearity, drift, and component tolerances. The second term is, of course, determined by the number of available quantization levels. Sometimes, when the accuracy is a function of where the measurement falls within the system's input range, the first term is stated in two parts:

$$\pm 0.01\% \text{ of FSR} + 0.01\% \text{ of Rdg}$$

where Rdg = reading

Caution should be exercised when evaluating accuracy specifications. They could either apply over a range of temperatures or be simply stated at room temperature (25° C).

4-3 PRECISION

If a system has difficulty in producing the same output code when making successive measurements of the same input value, it is said to have poor *precision*. Precision is therefore a measure of the system's repeatability. One cause of poor precision is internal system-generated noise. This specification is seldom found in the manufacturer's literature.

4-4 RESOLUTION/DYNAMIC RANGE

The ability of the data acquisition system to distinguish between adjacent analog input levels is called *resolution*. In a well designed system, the resolution is defined by the number of bits in the output code. For instance, a 10-bit system would have a resolution of 10 bits. In other words, the system can tell the difference between input levels that are 1/1,024 of the full-scale range apart (10 bits $= 2^{10} =$ 1,024).

Resolution also defines the system dynamic range. To explain, consider our 10-bit example and define the full-scale range to be 1.000 V. The smallest voltage that can be measured is then 1/1,024 V, and the dynamic range can be stated:

$$1{,}024{:}1 \qquad \text{or} \qquad \text{dB} = 20 \log_{10} 1{,}024 \cong 60 \text{ dB}$$

4-5 OVERRANGE

When the normal full-scale voltage is exceeded on many data acquisition systems, the output will continue to register the code assigned to the full-scale value. In some systems, however, a mode called "overrange" is provided for the over-voltage situation. While this feature allows the system to perform measurements beyond its normal range, care should be exercised, since accuracy is often degraded when in the overrange mode. Measurements of 120 to 130 percent of full scale are not uncommon.

4-6 LINEARITY

Two types of linearity specifications may be applied to the data acquisition system. The first, called *integral linearity*, is defined as the maximum deviation from a best straight line drawn through the transfer function curve (Figure 4-1). Integral linearity can be specified either as a percentage of full scale ($\pm 0.x$ percent of FSR) or in terms of a quantum level (less than $\pm\frac{1}{2}$LSB).

The second type of linearity is referred to as *differential linearity*. Differential linearity is a measure of the maximum deviation of any given quantum interval in the system's transfer function from its theo-

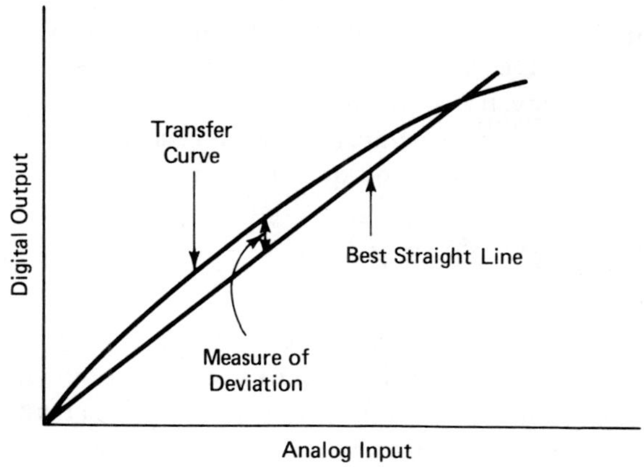

Figure 4-1. Integral Linearity

retical value. Examples of differential nonlinearity are displayed in Figure 4-2, in which the differential nonlinearity would be specified as ±½, the least significant bit (LSB). Differential linearity error may also be stated in terms of percentage of a quantum interval or LSB. The expression used is:

$$\% \text{ differential linearity} = \begin{cases} +\dfrac{(\Delta V_1 - \Delta V)}{\Delta V} \times 100\% \\ -\dfrac{\Delta V - \Delta V_2}{\Delta V} \times 100\% \end{cases}$$

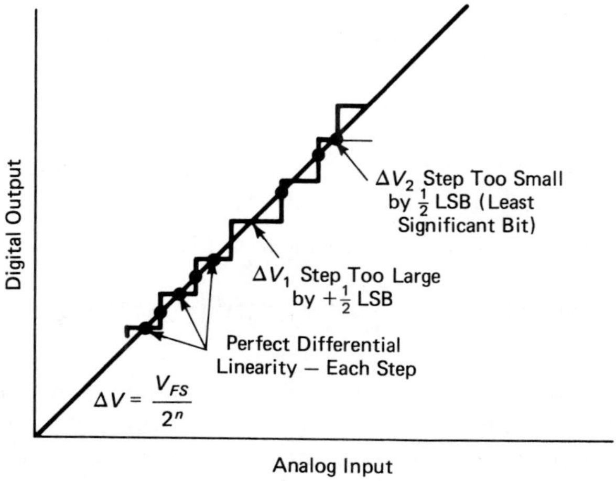

Figure 4-2. Examples of Differential Nonlinearity

For the example of Figure 4-2, the percentage of differential linearity would equal ±50 percent.

Differential linearity is an important specification when you have to sort things by means of analog input levels. Such situations may require very accurate bit sizes or quantum intervals.

4-7 MONOTONICITY

Monotonicity in a data acquisition system, or in any type of conversion hardware, is a guarantee that the system's output code will continue to increase with a continuously increasing input. It also ensures a decreasing output code for a continuously decreasing input. An example of a nonmonotonic transfer function appears in Figure 4-3. The most common place for this problem to occur in conversion systems is at the major decision points, namely, at the halfway and quarterway points of full-scale readings. Nonmonotonicity can be particularly troublesome if the acquisition system is part of a larger closed-loop control system.

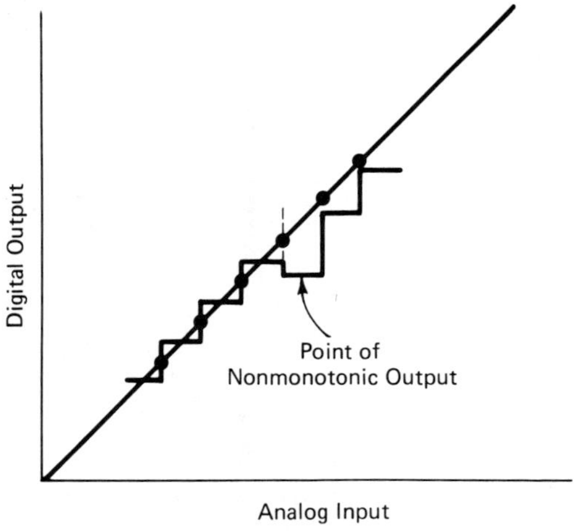

Figure 4-3. Example of Nonmonotonicity

4-8 TEMPERATURE COEFFICIENTS

Many of the parameters discussed in the previous sections are affected significantly by changes in ambient temperature. Two temperature effects that impair the system's ability to make accurate measurements —gain change and zero shift—are shown in Figure 4-4.

Temperature can also have adverse effects on integral and differ-

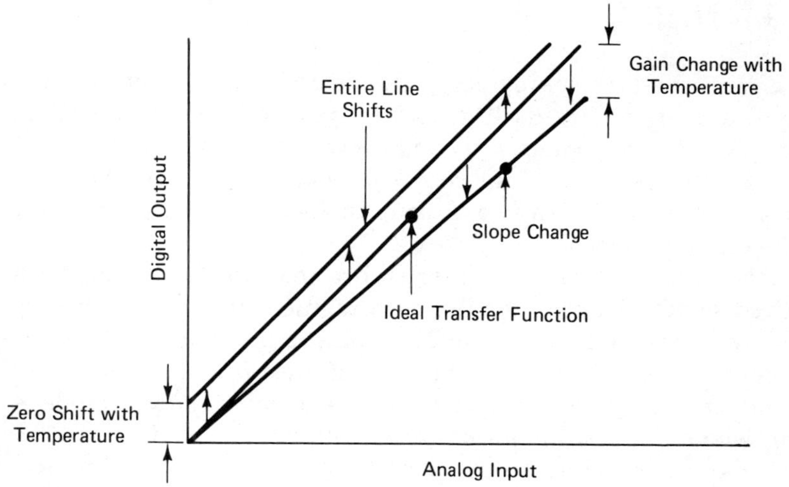

Figure 4-4. Temperature Effects on Accuracy

ential linearity and, if severe enough, can cause some systems to become nonmonotonic.

Temperature coefficients define the degree of inaccuracy that the temperature can cause in any parameter. Sometimes shortened to TEMPCOS, they are usually stated as a number of parts per million (ppm) per degree centigrade (/° C) of change. Some typical numbers would be:

Gain	±10 ppm/° C
Zero offset	± 5 ppm/° C
Integral linearity	± 5 ppm/° C
Differential linearity	± 2 ppm/° C

It is also common to have TEMPCOS stated in terms of the basic parameter, such as:

	Zero offset	±50 μV/° C
or	Accuracy	± 0.001%/° C

In any case, the temperature coefficients are factors that should not be overlooked. Too often systems have been designed with only the room temperature specifications in mind. Later use in the real environment has produced unsatisfactory results.

4-9 NOISE

Noise is a constant companion in the world of data acquisition. Even the measuring system itself generates systematic noise as it combines high-level digital signals with a low-level analog environment. If the discussion is restricted to the very front end of the system, however, we become concerned with two basic noise factors, namely normal mode and common mode.

Normal mode noise is interference that appears, along with the desired signal, between the two measurement terminals. This type of noise relies on the basic system band-pass response for filtering. Sometimes, filters are added at the various measurement inputs. Often manufacturers provide a specification known as normal-mode rejection, which defines the normal-mode filtering.

The other factor, *common-mode noise*, refers to interference that appears simultaneously on both measurement inputs. The majority of common-mode interference is attributable to the 60-Hz power frequency and its derivatives. The ability to reject this interference must be designed into the system's front end and is specified in the following fashion:

$$\text{Common-mode rejection (CMR)} = \begin{cases} 60 \text{ dB @ } 1 \text{ kHz} \\ 80 \text{ dB @ } 60 \text{ Hz} \end{cases}$$

For this specification to be useful, the drive impedance from the measurement source should also be known. Most CMR's are specified with a source impedance of 1 kΩ.

Common-mode rejection is discussed in greater detail in both Chapters 5 and 6.

4-10 CONVERSION TIME

The *conversion time* specification provides an indication of system sampling rate. It may be cited as the time required to complete a conversion, such as:

$$\text{Conversion time (12 bits)} = 50\ \mu\text{sec}$$

or as a rate, such as:

$$\text{Conversion rate (12 bits)} = 20 \text{ kHz}$$

Where the number of bits utilized is optional, it is not uncommon to state conversion rate in terms of time/bit, as follows:

$$\text{Conversion time} = 10\ \mu\text{sec/bit}$$

If a given system were capable of 10-bit operation, then it requires 100 μsec for a total capability conversion. Should only 6 bits be required, the conversion time would be 60 μsec.

An acquisition system that is aimed particularly at communications may have its conversion time stated in terms of bits per second. In this case, the conversion rate is the product of the conversion time and the number of bits of resolution. For example:

Conversion time (10 bits) = 1 μsec
Conversion rate = 10 Mbits/sec

Often associated with conversion hardware is a parameter called *conversion delay*, which is simply a "black box" delay. It denotes the time required from sampling of the input to useful system output. In general, once the first conversion output is available, successive outputs occur at the system conversion rate.

4-11 SETTLING TIME

The *settling time* of a system refers to the time elapsed between the application of a full-scale input step and the stabilization of the output within a specified error band. It is a parameter that applies to the entire analog portion of the data acquisition system up to the point of sampling.

Figure 4-5 is an example of the settling time situation. Once the overshoot has settled to within the error band, an accurate conversion is possible. This concept is not restricted to analog-to-digital conversion of step inputs or pulse tops. A similar situation exists when the analog multiplexer switches from a channel whose input is at the low

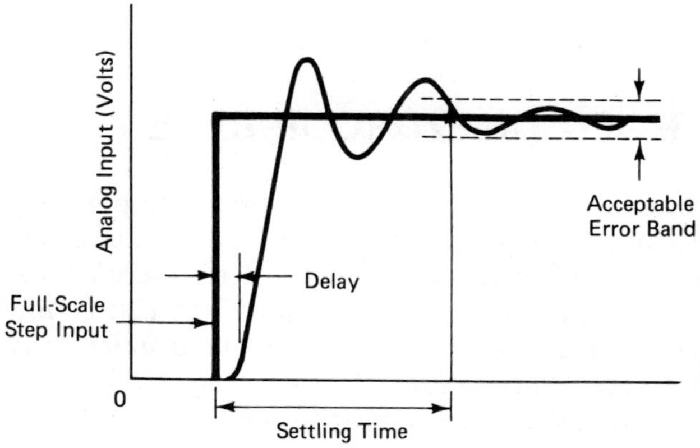

Figure 4-5. Example of Settling Time

end of the input range to a channel whose input is at the high end of the input range. The settling time specification is most often stated as:

settles to within $+0.01\%$ in 20 μsec.

4-12 APERTURE TIME

The width of the sampling window, or the actual time the system requires to obtain a sample of the analog input, is called *aperture time*. It is also often referred to as the "uncertainty of sampling." Hence another term often used is *aperture uncertainty*. (Mention of the aperture time parameter is included in this section for completeness; a more comprehensive treatment is included in Section 2-6.)

If the sampling function is highly repeatable and the sample window small with respect to the period of the highest frequency being sampled, then the effective aperture may be defined by the small time uncertainty in that sample function. This is an important concept because an order of magnitude improvement can be realized in effective aperture time. For example, a system may have an aperture time of 50 nsec, while aperture uncertainty is only 5 nsec.

4-13 ACQUISITION TIME

Acquisition time is a characteristic that applies to systems containing a sample-and-hold function. It defines the time required to go from one "hold" situation to a new one, within the specified system accuracy. The specification covers the following worst-case situation: The voltage held is at one end of the system's full-scale input value, and at the next sample it must acquire a voltage that is at the other end of the input value scale. A typical example is:

Acquisition time to 0.1% of full scale $=$ 4 μsec

4-14 HOLD TIME/DROOP RATE

Hold time and *droop rate*, also sample-and-hold parameters, are different ways of referring to the same mechanism. The sample-and-hold operation is based on the acquiring of a charge on a capacitor and maintaining it until an accurate measurement can be made. A measure of how well the system performs the hold function is defined by either hold time or droop rate.

Hold time defines how long the system can hold the stored value to a given accuracy, while *droop rate* indicates the loss of charge per unit of time. The specifications for each might be stated as:

Hold time to within 0.1% = 5 msec
Droop rate = 50 mV/sec

In the first case, the system will hold the voltage to within a measurement accuracy of 0.1 percent for a period of 5 msec. For the droop rate situation, the voltage on the hold capacitor is decaying at a rate of 50 mV per second.

Although both specifications are still used, droop rate is more popular because it allows users a ready way to calculate results for their particular situations.

4-15 CROSSTALK

A particularly important parameter in a multichannel data acquisition system is *crosstalk*, which provides an indication of signal leakage from one channel to any other. Crosstalk is a function of capacitance between adjacent channels, of the source impedance of the affected channel, and of the frequency of the interfering signal or signals. A typical crosstalk specification is:

Channel crosstalk = −80 dB at 10 kHz
(Channel source R = 1 kΩ)

4-16 dc INPUT CHARACTERISTICS

The most overlooked data acquisition system characteristics are those relating to the input. It is easy for the user to forget that offset voltages, offset currents, and bias currents are associated with these inputs. Consider a system where the input amplifier presents a bias current of 1 μA. The input is also presented with a source resistance of 10 kΩ.

The equivalent circuit appears in Figure 4-6. The total offset is:

$$V_{\text{Offset}} = I_{\text{bias}}(R_S) = (10^4)(10^{-6}) = 10 \text{ mV}$$

An offset of 10 mV is enough to shift many system inputs by several least significant bits. To compound the problem, a multichannel system could produce a different value of offset on each channel. The user must take this into account when designing, using, or calibrating the system. Fortunately, the advent of field-effect transistor (FET) inputs has substantially reduced this problem.

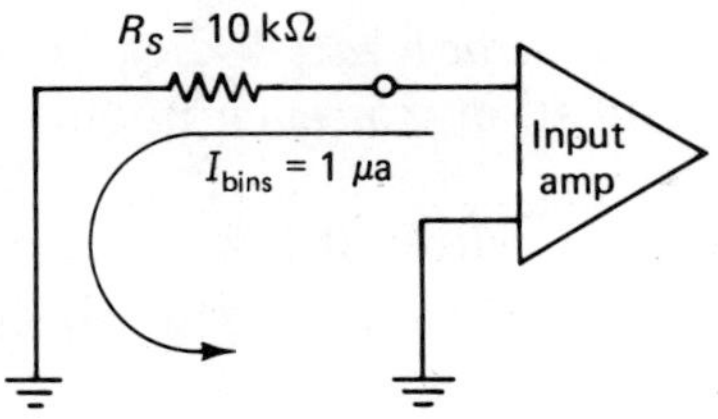

Figure 4-6. dc Input Characteristics

PROBLEMS

4–1. A data acquisition system requires a dynamic range of 72 dB and an overall analog accuracy of ±0.5 percent at room temperature.
a. How many output bits are required?
b. How would the accuracy be specified?

4–2. A data acquisition system with a full-scale voltage of 10,000 V has an input nonlinearity function as shown in Figure 4–1. At the worst-case point, the deviation is +2.4 mV. What is the maximum number of bits of resolution that the system can produce while keeping the nonlinearity less than ½LSB?

4–3. A quantizer has a differential linearity of ±25 percent and a theoretical quantum interval value of 10 mV. What are the minimum and maximum quantum step sizes for this converter?

4–4. The accuracy specification for an acquisition system with a 10-V full-scale input and the input offset zero's is:

±0.1% of FS ± ½LSB @ 25° C

The system TEMPCOS are given by the following two parameters:

Zero offset:	±50 μV/° C
Accuracy speed:	± 0.001%/° C

Restate the system accuracy at 50° C.

4–5. An analog-to-digital converter has a conversion *time* specification of 10 bits at 5 μsec.
a. How would the conversion *rate* be specified?
b. If the converter utilizes a successive approximation architecture where conversion takes place one bit at a time, how might the conversion time be specified?

4–6. The hold time and droop rate example specifications of Section 4-14 are not for the same sample-and-hold system. Consider V_{FS} to be 10,000 V for both cases and, for the second, consider the hold time to be 100 msec. Calculate the droop rate for the first case and the hold time for the second.

4–7. A data acquisition system input has a circuit similar to that shown in Figure 4-6. In this case, however, $R_S = 5\ k\Omega$, $I_B =$ 100 nA, and $V_O = +4$ mV. The full-scale input range is +10 V, and the conversion resolution is 10 bits.

a. If the inaccuracy allowed for the input is ±½LSB, is this input satisfactory?

b. If both I_B and V_O had TEMPCOS of +0.2 percent/° C and ΔT was +25° C, would the input still be satisfactory?

REFERENCES

1. D. H. Shiengold, *Analog–Digital Conversion Handbook.* Norwood, Mass.: Analog Devices, 1972.
2. Datel Systems, Inc., *Engineering Product Handbook.* Mansfield, Mass.: Datel Systems, Inc., 1977–1978.
3. B. M. Gordon, "What's Wrong with A/D Converter Specs," *EEE Magazine*, Vol. 17, No. 2 (February 1969).
4. B. M. Gordon, "Digital Sampling and Recovery of Analog Signals," *EEE Magazine* (May 1970).

5

Analog Multiplexers

5-0 INTRODUCTION

The analog multiplexer allows the use of a single, highly sophisticated analog-to-digital conversion system for the measurement of a number of different signal sources. Although most multiplexer systems accommodate from 10 to 100 channels, systems having up to 1,000 channels have been implemented.

In Figure 5-1, you can find the analog multiplexer depicted in its most basic form. Each of the analog signals are individually switched to the input of the conversion system. In this manner, the signal conditioners and the analog-to-digital converter, which follow, are time-shared with the various input signals. The major premise is that the required sample rate for each of the signals be satisfied. Later in the chapter we will find that it is not uncommon to find as many as three switches per signal, or analog channel, in high noise environments.

A top-level block diagram of the typical analog multiplexer system can be seen in Figure 5-2. The system consists of three basic parts:

1. the analog switches,
2. the individual switch drivers, and
3. the control logic.

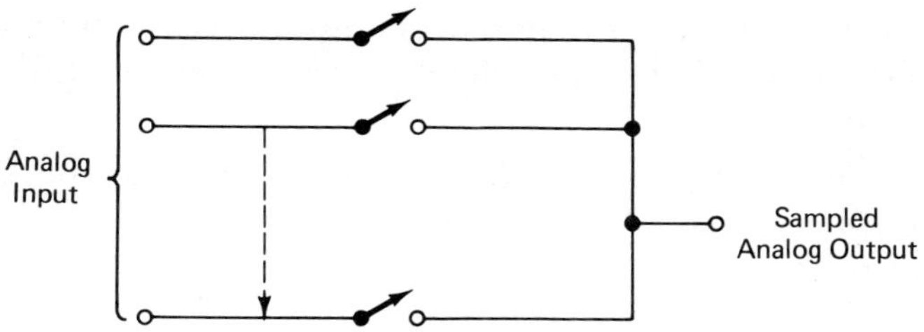

Figure 5-1. The Basic Analog Multiplexer Switch

While the drivers provide the signals to turn the switches on and off, the control logic determines the order in which the switch channels are sequenced.

There are three general modes of multiplexer operation:

1. In the *sequential mode*, each channel is activated in some continuous predetermined sequence. The sequence could be as simple as starting with Channel 1 and proceeding in order to Channel *n*. Yet channels that require higher sample rates could also appear more often than others in the sequence. Whatever the sequence, the system repeats it on a continuous basis as long as it is active.
2. In the *random-access mode*, any channel can be selected individually by presenting the channel address at the random-access

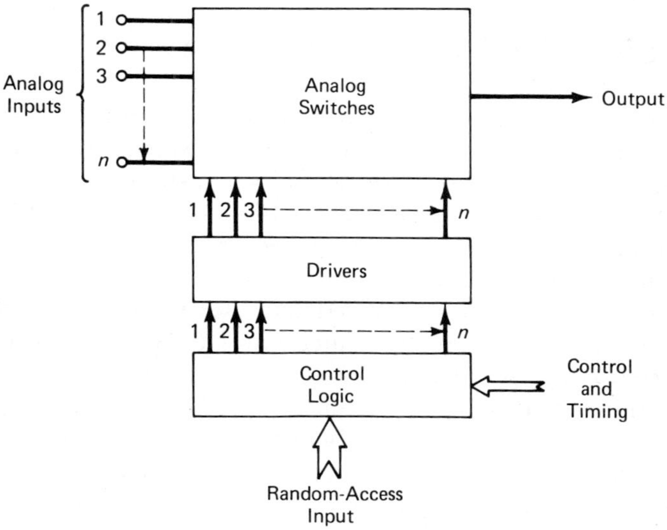

Figure 5-2. Analog Multiplexer System Block Diagram

input, with the addressing scheme usually based on some binary sequence. This mode is generally used when the DAS is under computer control.

3. When in the *manual mode*, the multiplexer can be advanced from channel to channel by some type of manual selection on the front of the system.

Some multiplexer systems offer all three. Greater detail on the implementation of the various multiplexer modes can be found in Chapter 10.

There is, of course, an alternative to analog multiplexing. You could provide signal conditioning and A/D conversion for each channel and perform the multiplexing digitally. Until recently, such a system was not economically feasible, even for a few analog channels. As conversion hardware continues to decrease in price, however, the digital multiplex system becomes increasingly attractive for many DAS problems. Such an alternative has another distinct advantage in that each channel can be individually tailored to the signal being measured. Yet many situations still exist in which the analog multiplexer provides savings in power, weight, size, and cost, and the need for such a device seems far from ending.

The remainder of this chapter provides an insight into the types of devices used for switches, as well as the various multiplexer configurations. It also considers the important operational parameters and characteristics essential for development and use.

5-1 SWITCH TYPES

Analog multiplexer switches may be separated into two broad categories —electromechanical and semiconductor. In general, electromechanical devices are used for low-level signals when speed is not a problem, while semiconductor devices are utilized for relatively high-level, high-speed situations. While the electromechanical device more closely approximates a perfect switch (zero "on" resistance and infinite "off" resistance), its relatively slow speed relegates it to under a thousand operations per second. Semiconductor diode switches, on the other hand, can operate in the megachannel-per-second range. The penalty paid for this speed, however, is substantial "on" resistance and finite "off" resistance, along with a number of other undesirables such as bias voltages, leakage currents, and interchannel crosstalk. Either type of device has its place, is widely used, and is worthy of discussion.

5-1.1 The Stepping Switch

Perhaps the most rudimentary of electromechanical devices is the stepping switch multiplexer. A series of high conductance contacts, each connected to an individual analog input, is arranged in a circular pattern, as shown in Figure 5-3. A separate contact, moving in a circular fashion, connects first one and then the next analog input to the system measuring point (analog output). The movable contact is driven by a stepping motor, which moves the contact from input to input at rates between one sample every two seconds to one sample per second. Thousands of this type of multiplexer have been sold as *multipoint* recorders by the instrumentation industry. These units, which accommodate either 24 or 48 channels, are portable, rugged, and relatively inexpensive.

They have one major drawback, however. In the basic unit, the data is recorded by placing an ink dot on a sheet of moving graph paper for each channel, as it is sampled. The output is a set of interwoven data that must be unscrambled manually. In recent years, the multipoint recorder has been coupled with a classical DAS to form a multipoint recorder logger. As each channel is sampled, its value is A/D converted and the results stored on some medium, such as punched paper tape.

The major characteristics of the stepping switch multiplexer are:

Rate:	1 sample every 1 to 2 seconds
"On" resistance:	20 to 200 mΩ
"Off" resistance:	100 MΩ minimum @ 50 V
Life:	1,000 hours operating time (contacts increase in resistance and must be cleaned)
Number of channels:	24 or 48

One other problem with this type of multiplexer is that it can operate only in a sequential manner—where one numbered channel must follow the other. In some units, channels may be skipped, but those used must be taken in consecutive order. To increase the sample rate per channel, a given signal must be tied to more than one stepping switch input.

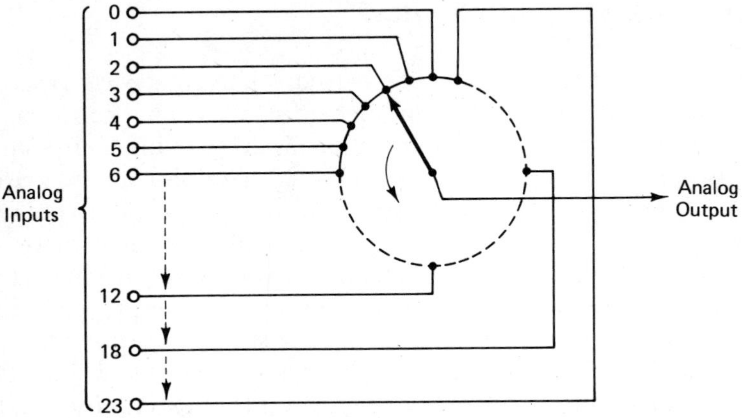

Figure 5-3. The Stepping Switch Multiplexer

5-1.2 The Motor-Driven Commutator

The motor-driven commutator is similar to the stepping switch multiplexer in that the channels are sampled in consecutive order. It does, however, operate at a much higher speed and can accommodate many more channels. A pictorial of a typical motor-driven commutator can be seen in Figure 5-4.

The analog inputs are connected to contacts located on the periphery of a circular insulator. The inputs are connected to the common analog output point by means of a motor-driven arm, which consecutively joins each contact to a central conducting ring. Units are available with as many as 180 input contacts, and with speeds as high as 180 channels per second.

This type of multiplexer allows the use of multiple switches per channel by adding two or three concentric rings of contacts on the insulator, by adding additional concentric conducting rings, and additional bars to the motor-driven arm. The major characteristics of the motor-driven commutator are as follows:

Rate:	Up to 180 channels/second
"On" resistance:	Less than 1 Ω
"Off" resistance:	100 MΩ @ 50 V
Life:	1,000 hours of operation before contact degradation
Number of channels:	Up to 180 per contact ring; up to 4 contact rings
Contact rating:	10 mA @ 10 V

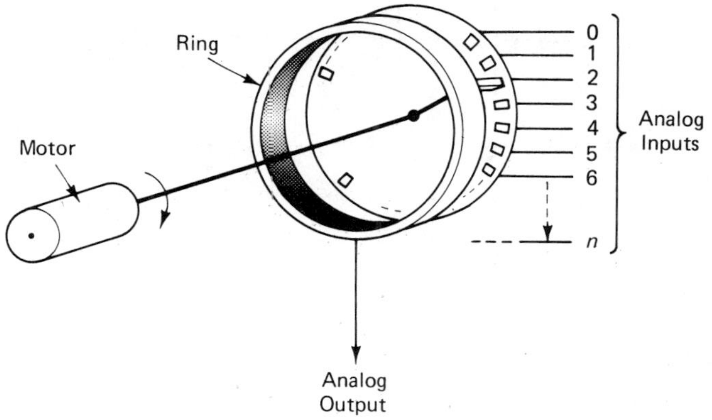

Figure 5-4. Motor-Driven Commutator

The major drawback of this device, like the stepping switch, is that it does not allow programming and must proceed from channel zero to channel n and so on. Again, higher sample rates may be achieved only by connecting a given analog input to more than one commutator input.

5-1.3 Reed Relays

Perhaps the most popular electromechanical multiplex device is the reed relay. Although most of those used are dry reeds, mercury-wetted reeds are found where higher contact current ratings and/or longer life is desired.

Dry Reeds

A pictorial of the dry reed relay can be found in Figure 5-5. It consists of a pair of flat reed contacts enclosed in a glass envelope, which is filled with an inert gas such as dry nitrogen. The contacts, which are made of a precious metal such as gold, rhodium, or silver, do not degrade because of the inert atmosphere. The reeds are mechanically biased open. When a current is passed through the relay coil, one reed is magnetized north and the other is magnetized south. They therefore attract and close.

The drive circuit for the reed relay is shown in Figure 5-6. It is a simple, one-transistor circuit, connected to the opposite side of the relay coil from the supply voltage. When the switch is activated, the current through the coil causes the relay contacts to close. The suppression diode shorts out the back-EMF effects, which occur when the

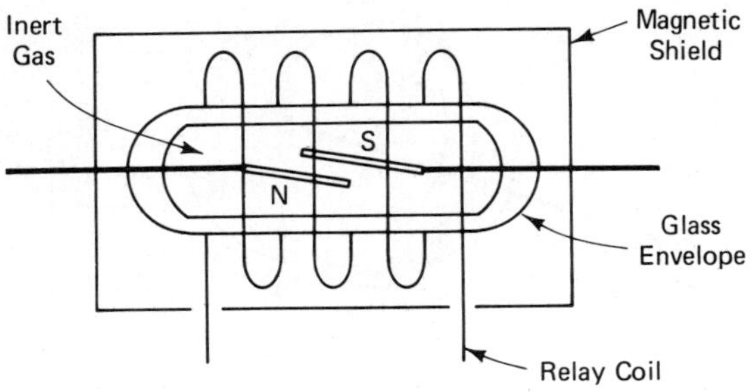

Figure 5-5. Dry-Reed Relay

switch is opened. Relays that operate with 6, 12, or 24 V are available.

Major features of the reed relay are its low contact or "on" resistance (typically 10 to 20 mΩ) and reasonable speed (200 to 500 channels/second). Because of the extremely low "on" resistance, the device is used for very low-level multiplexing. So the switch is usually supplied with a magnetic shield. The shield may be grounded to eliminate unwanted stray noise from coupling in to the contacts and from contaminating the signal being multiplexed. Characteristics of the reed relay multiplex switch are:

Rate:	200 to 500 channels/second
Actuate time:	750 to 1,000 μsec
Release time:	50 to 700 μsec
"On" resistance:	10 to 150 mΩ
"Off" resistance:	500 K MΩ @ 200 V
Contact rating:	1 mA @ 10 V
Life:	>1 billion operations
Switching noise:	100 μV into 100 kΩ initially; 1 μV—50 μsec after closure

The initial higher noise level is due to switch bounce, which is characteristic of dry reed devices.

Mercury-Wetted Reeds

The mercury-wetted reed relay can be seen in Figure 5-7. Its operation is identical to that of the dry reed with one exception. The mercury in the bottom of the glass envelope wicks up onto the lower contact and provides a layer between the two contacts when they are closed. This eliminates contact bounce and thus wear and tear on the contacts. It also allows higher contact current ratings. Due to the same wicking

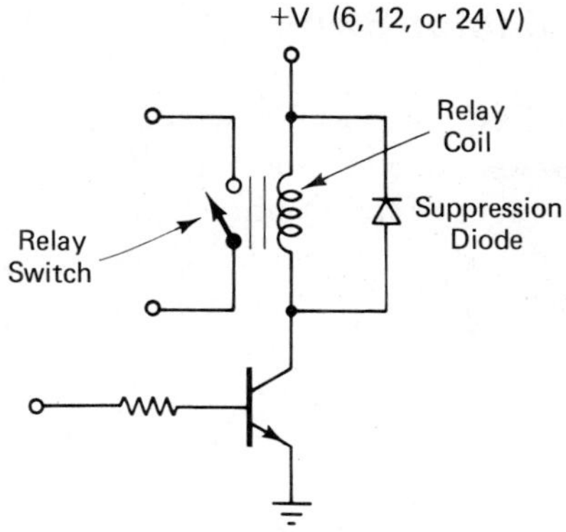

Figure 5-6. Typical Reed Relay Drive Circuit

action, the mercury-wetted reed is much slower than the dry reed. It also has a serious drawback in that the envelope is position-sensitive; that is, it must always be operated in the upright position. Major characteristics of the wet reed switch are as follows:

Rate:	100 channels/second
Operate time:	2.0 msec for both actuate and release
"On" resistance:	20 to 50 mΩ
"Off" resistance:	500 K MΩ @ 200 Ω
Contact rating:	10 mA @ 10 V
Life:	>20 billion operations
Switching noise:	275 μV into 100 kΩ initially; 10 μV—4 msec after closure.

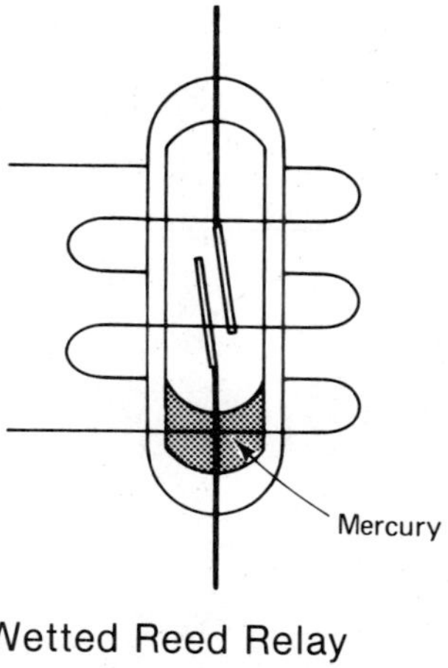

Figure 5-7. Mercury-Wetted Reed Relay

5-1.4 Semiconductor Switches

Semiconductor switches have a far wider range of application in today's data acquisition system than do electromechanical devices. They offer the advantages of being smaller, lighter, and more reliable. They may also be operated at much higher speeds while providing reasonable accuracy. They require much lower level drive circuitry and have no contact bounce. Some of the disadvantages are that they act as resistive contacts when on and provide only fair isolation when off. They also have leakage currents and some finite offset voltage. Unlike the electromechanical switch, poor isolation exists between signal path and drive voltage. The most popular semiconductor devices are fast diodes, junction field-effect transistors, P- and N-MOS, and CMOS devices. A brief discussion of each follows.

Diode Bridge Switch

The diodes used in diode bridge switches are Shottky barrier or hot-carrier type devices, which are characterized by extremely fast off-to-on switching and very fast reverse recovery times, when turning off (Figure 5-8). Such switches, which can switch in a few hundred pico-seconds, are limited more by circuit physical construction than by the diodes themselves. It is not uncommon for hot-carrier diode quad multiplexers to operate at rates in excess of 10 MHz. Looking at Figure 5-8, we see that, when the switch is enabled, all four diodes are on and

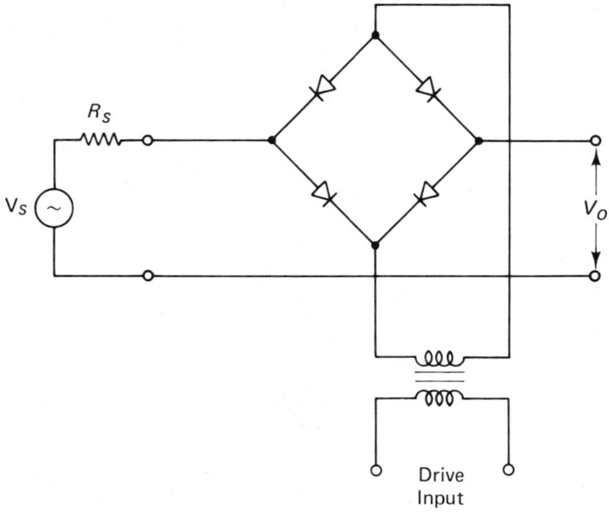

Figure 5-8. Diode Bridge

the "on" resistance is that of a single diode. The transformer provides isolation between the drive circuitry and the signal being multiplexed. Characteristics of this switch (typical at 25° C) are:

Rate:	10 megachannels/second
"On" resistance:	5 to 25 Ω
"Off" resistance:	1 MΩ, minimum
Operating noise:	5 μV, typical
Offset voltage:	20 to 50 μV
Input range:	±5 mV to ±10 V
Reliability:	15,000 hours MTBF

Junction Field-Effect Transistors (J-FETS)

The junction field-effect transistor is a very fast, low "on" resistance multiplexer switch. Its major disadvantage is that it is seldom found in a multiswitch array configuration. A cross-section of a typical N-channel device can be seen in Figure 5-9.

Without bias, the path from the source to drain is completed, while the N material simply acts as a small value resistor (5 to 50 Ω). When a gate-to-source voltage of the polarity shown is applied, a field is established that "pinches" off the channel. This points out a peculiar problem with the J-FET (depletion mode) switch. When the multiplexer power is off, all channels are turned on. This state connects all channels and therefore all the analog signal sources together. In some cases, such a situation could prove disastrous, and suitable precautions must be taken. As in all semiconductor switches, the channel "on" resistance is temperature-sensitive and increases approximately 0.5 percent per degree of Centigrade. It has a very small leakage current

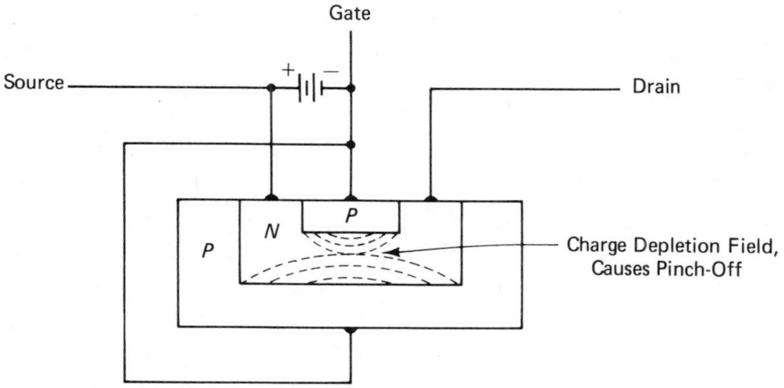

Figure 5-9. N-Channel Junction Field-Effect Transistor

(50 pA to 1 nA), which doubles for every 10° C rise in temperature. Typical J-FET characteristics (typical at 25° C) are:

Rate:	>1 megachannel/second
"On" resistance:	5 to 50 Ω
"Off" resistance:	10^{10} Ω (1 nA leakage at 10 V)
Operating noise:	<5 μV
Offset voltage:	N/A
Input range:	±5 mV to ±10 V
Reliability:	10,000 hours MTBF

P- and N-MOS Transistors

Another popular field effect device used for analog multiplexing are P-and N-MOS (metal oxide semiconductor) transistors. Unlike the depletion mode J-FET devices, the MOS transistors are enhancement mode and therefore require gate signals to turn them on. Another advantage of this type of switch is that it is readily found in arrays of up to 8 channels. On the negative side, they are generally slower and have a much higher "on" resistance than the J-FET.

A cross-section of a typical P-MOS switch can be seen in Figure 5-10. With 0 V between the metal gate (which is insulated from the substrate by an oxide layer) and the source, the device is off. When a negative gate-to-source voltage is applied, positive charges accumulate at the oxide/semiconductor interface. Eventually the point of threshold will be reached where enough charges have accumulated to effect a channel between source and drain. The switch is now on. In the "off" condition, the source-to-drain path looks like two diodes connected back-to-back, thereby providing good isolation, low leakage, and low

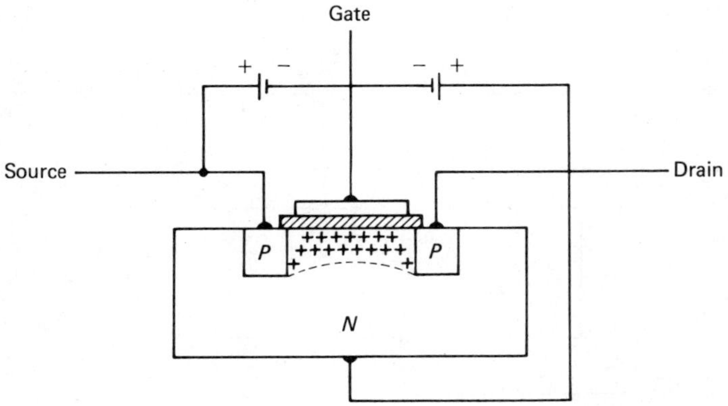

Figure 5-10. P-MOS Transistor

offset voltages. Like all semiconductor devices, the "on" resistance is temperature-dependent. In the MOS device, however, the "on" resistance is also a function of the body (or substrate) to source voltage. When using the P-MOS device, you must take care always to keep the substrate more positive than the source or drain. Operation of the N-MOS device is identical to that of the P-MOS device, with the polarities reversed. Characteristics of P- or N-MOS analog switches (typical at 25° C) are:

Rate:	>100 kilochannels/second
"On" resistance:	100 to 1,000 Ω
"Off" resistance:	10^{10} Ω (1 nA leakage @ 10 V)
Operating noise:	<5 μV
Offset voltage:	<5 μV, typically
Input range:	±5 mV to ±10 V
Reliability:	>10,000 hours MTBF

CMOS Switch

In recent years, the CMOS (complementary metal oxide semiconductor) switch has become popular as an analog multiplexing device. Besides being well suited to multiple switch integration in a monolithic array, similar to P- or N-MOS, it offers two other distinct advantages. First, the complementary circuitry does not require the constant, heavy power drain of the standard MOS system. It consumes power only at switch transition times when capacitance must be charged or discharged. Second, the basic CMOS switch circuit (see Figure 5-11) contains both a P-MOS and an N-MOS device, connected in a parallel arrangement. As the figure shows, both P- and N-channel devices are integrated into the common bulk substrate. Operation of the individual devices is identical to that described in the previous section on P- and N-MOS switches.

Yet there is one distinct and significant advantage. The parallel connection provides an "on" resistance that is relatively constant with source-to-gate voltage. As Figure 5-11c shows, when the P-channel device is turning off, the N-channel device is turning on—and vice versa. This setup provides a switch that is reasonably independent of input signal voltage.

Some of the early circuits had a problem with latchup when the input voltage exceeded the substrate voltage. Special circuit techniques have all but eliminated this problem.

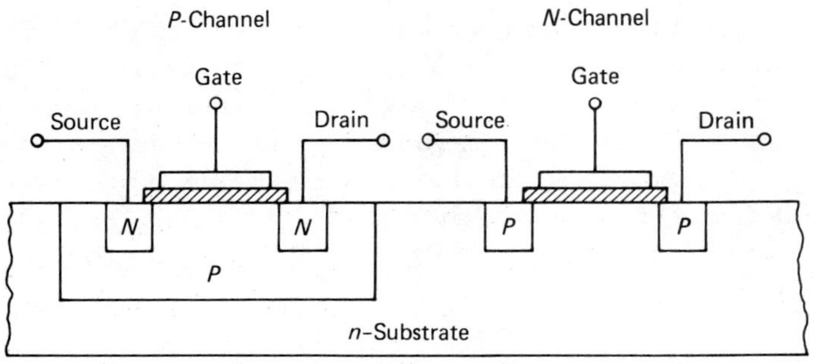

(a) Physical representation

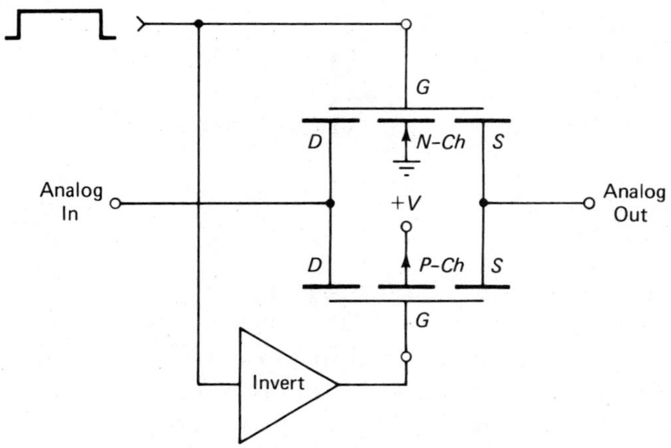

(b) Typical circuit connection

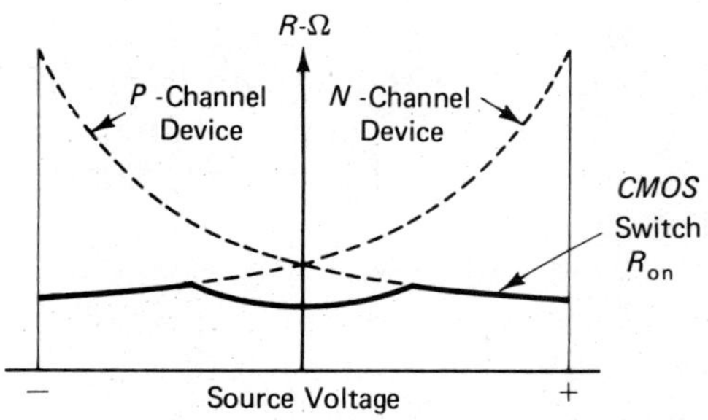

(c) "On" resistance vs source voltage

Figure 5-11. The CMOS Switch

The basic CMOS switch characteristics (typical at +25° C) are:

Rate:	>100 kilochannels/second
"On" resistance:	50 to 800 Ω
"Off" resistance:	≈10^9 Ω (10 nA leakage @ 10 V)
Operating noise:	<5 μV
Offset voltage:	5 μV
Input range:	±5 mV to ±10 V
Reliability:	>10,000 hours MTBF

5-2 CIRCUIT CHARACTERISTICS

The basic analog multiplexer equivalent circuit can be seen in Figure 5-12. Just as you would expect, the circuit is heavily influenced by the analog input source parametrics, as well as by the input characteristics of the circuitry that follows. As we shall see, the ideal system is driven from a very low source impedance into a very high load impedance. When that situation exists, both static and dynamic problems are reduced to a minimum. Although the circuit in Figure 5-12 is simplified, it does present the most significant parametric components of an operational multiplexer system. In this section, we will discuss the characteristics that most influence multiplexer operation, with the circuit in Figure 5-12 used as the model.

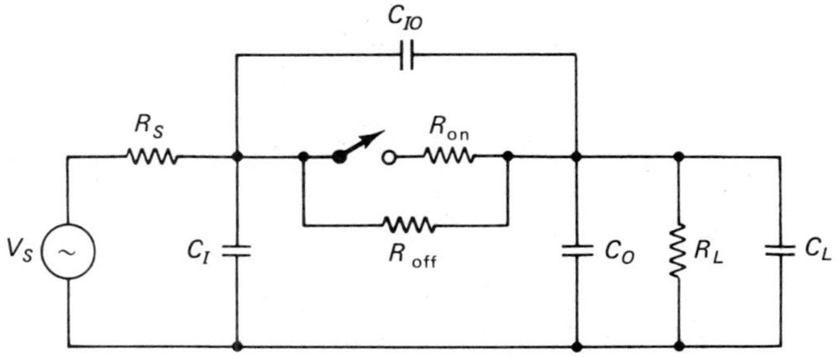

Where: R_S = source resistance; C_I = $C_{in\,(switch)} + C_{source}$; R_{on} = switch - "on" - resistance; R_{off} = "off" - leakage path; C_{IO} = coupling C, across open switch; C_O = output C, all switches; R_L = load resistance; C_L = load capacitance

Figure 5-12. The Basic Switch Equivalent Circuit

5-2.1 Leakage

Leakage is a problem that particularly plagues semiconductor multiplexers. As we have seen, semiconductor devices all contain parasitic diodes formed by their various N- and P-doped materials. These diodes are always maintained in a back-biased condition during normal operation. We know, of course, that leakage currents are associated with back-biased diodes. This leakage current, whose value doubles for a 10° C rise in temperature, causes a problem.

An equivalent circuit for leakage current can be found in Figure 5-13. Since this is a static problem, all of the parasitic capacitances have vanished from the model. The model has also been reoriented so that the "on" switch and its associated source resistance effectively shunt the input resistance of the load. (Note: They also short all the "off" resistances of the other switches.) As R_L is generally very much greater than $R_S + R_{on}$, the offset error at the measuring point is:

$$e_{OE} = (n - 1)I_L(R_S + R_{on}) \qquad (5\text{–}1)$$

where n = number of multiplexed channels

I_L = leakage current of a single "off" switch

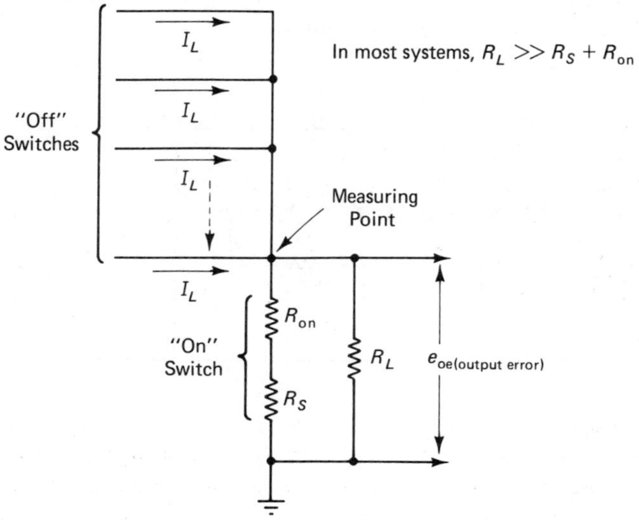

Figure 5-13. Equivalent Circuit for Leakage Current

Given an example where:

$$R_S = 1{,}000\ \Omega \qquad R_L = 10\ \text{M}\Omega$$
$$R_{on} = 200\ \Omega \qquad n = 15$$
$$I_L = 2na\ @\ 25°\ \text{C}$$
$$e_{OE} = 14(2 \times 10^{-9})(1.2 \times 10^{3}) = 33.6 \times 10^{-6}\ \text{V}$$

for a 12-bit system with a full-scale value of 100 mV, each quantum level is 24.4×10^{-6} V. The offset would be more than a quantum level in this system. Fortunately, it is the difference between the offsets produced by both channels that is important, since the nominal value can be adjusted out at a common point in the DAS.

This problem becomes more severe as the number of channels increases and/or as the source resistances vary widely. One technique for improving this situation can be seen in Figure 5-14. Here the multiplexer is split into two "tiers," separated by a second level of multiplexing. For the configuration shown, an improvement of nearly three to one is realized, as follows:

	*3*n *Single Channels*	*3*n *Multitiered*
I @ output	$(3n - 1)I_L$	$(n + 1)I_L$

Many more segments could be implemented, and it would not be uncommon to see a 100-channel multiplexer constructed as 10 groups of 10 switches.

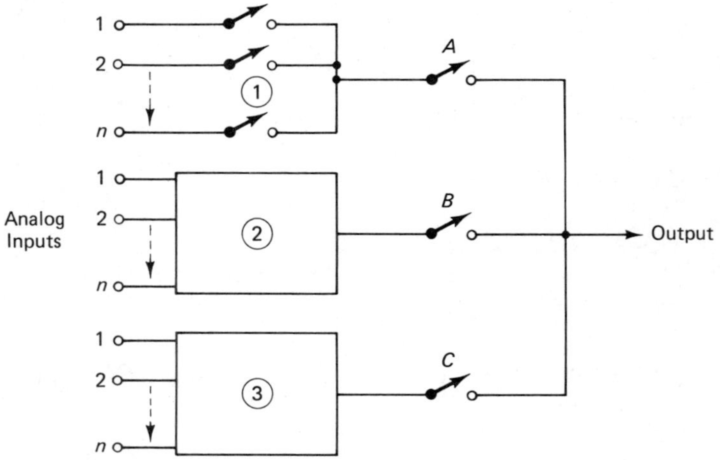

Figure 5-14. Multitiered Multiplexing

5-2.2 ac Response

There are two areas of interest when considering the ac response of an analog multiplexer. The first and most important is the time required to switch from one channel to another and to settle out to the required system accuracy, while the other concerns the system bandwidth once the switch is closed. Although you are, perhaps, most familiar with the first, you cannot ignore the second, since it is fundamental to the operation of many pieces of signal and spectrum analysis equipment.

Both situations depend on the equivalent circuit of Figure 5-15a. The parameter C_{OT}, used in Figure 5-15a, represents the parallel combination of the output capacitances, C_O, of all the switches connected to the measuring point. Even though this circuit is already a simplified version of the total multiplexer, much can be gained by simplifying further to the single time constant circuits of Figure 5-15b and c. These simple circuits provide a great insight into the multiplexer operating speed, even if they do not pinpoint it exactly. For instance, if the simple circuit indicates a 20- to 50-percent speed margin, it is highly likely that the multiplexer will satisfy the system requirements. On the other hand, if the circuit shows a two-to-one speed deficiency, you can rest assured that no amount of detailed analysis will make up the difference.

Considering the settling time problem first, we find the following time constant (TC) expressions for each of the circuits of Figure 5-15:

For 5-15a ($R_S \approx R_{\text{on}}$, and $C_I \ll C_T$):

$$\text{TC} = (R_S + R_{\text{on}})C_T \tag{5-2}$$

For 5-15b ($R_S \ll R_{\text{on}}$):

$$\text{TC} = R_{\text{on}}C_T \tag{5-3}$$

For 5-15c ($R_{\text{on}} \ll R_S$):

$$\text{TC} = R_S(C_I + C_T) \tag{5-4}$$

You can also easily prove that in terms of a single RC time constant that:

$$t_s = \text{RC} \ln \frac{100}{\%\ \text{error}} \tag{5-5}$$

where t_s = settle time

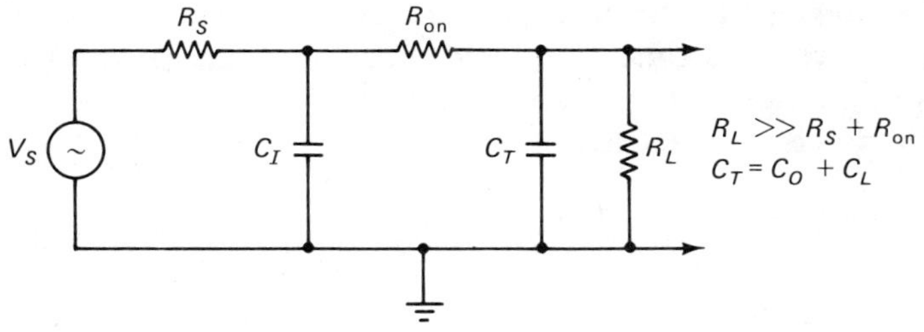

(a) Total equivalent circuit

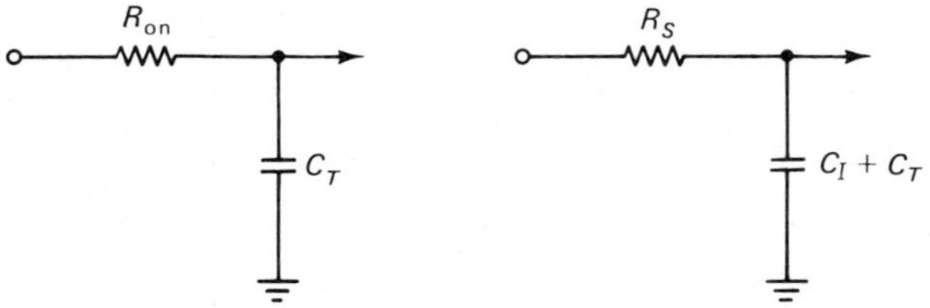

(b) For $R_S << R_{on}$ (c) For $R_{on} << R_S$

Figure 5-15. Equivalent Circuits for ac Response

Consider the following example:

$R_{on} = 10\ \Omega$		$C_{OT} = 100$ pF	
$C_L = 20$ pF		$R_L = 10\ \text{M}\Omega$	
$C_I = 5$ pF		Accuracy required: 0.1%	

With $\mathbf{R_S = 0}$ (select circuit 5-15b):

$$t_s = 1{,}200 \times 10^{-12} \ln \frac{100}{0.1} = 1.2 \times 10^{-9} \ln 1{,}000$$

$$t_s = (1.2 \times 10^{-9})(6.9) = 8.29 \times 10^{-9} \text{ sec} = 8.29 \text{ nsec}$$

With $\mathbf{R_S = 1{,}000\ \Omega}$ (select circuit 5-15c):

$$t_s = (10^3)(125 \times 10^{-12}) \ln 1{,}000$$

$$t_s = (1.25 \times 10^{-7})(6.9) = 8.625 \times 10^{-7} \text{ sec} = 863 \text{ nsec}$$

These examples clearly show that the source resistance has a dramatic effect on the switching time of an analog multiplexer—the lower the resistance, the faster the multiplexer. Unfortunately, for economic reasons, you cannot always choose a solution that reduces the inherent high output resistance of many transducers. The multitiered system shown in Figure 5-14 also improves the ac response because it effectively reduces the capacitance seen at the measuring point. For the configuration shown, an improvement of nearly three to one in speed may also be realized, as follows:

	*3*n *Single Channels*	*3*n *Multitiered*
Effective output capacitance (C_{OE})	$3n\ C_{OS}$ (capacitance at out switch)	$(n + 3)\ C_{OS}$

Unfortunately, the settling time problem is not this simple for a fast semiconductor multiplexer. A switching spike or transient caused by driver circuit feed-through from gate to device output is often found superimposed on the RC component. These fast transients cause ringing at the switch output. The ringing must also dissipate before switch settling is complete. This phenomenon is a function of the gate-signal, the multiplex switch type, and multiplexer physical construction.

The second ac response factor of concern is the bandwidth of the closed switch. The models of Figure 5-15 also apply here. For the three circuits, the bandwidth expressions are:

For 5-15a ($R_S \approx R_{\text{on}}, C_T \gg C_I$)

$$f_{3\text{dB}} = \frac{1}{2\pi(R_S + R_{\text{on}})C_T} \tag{5-6}$$

For 5-15b ($R_S \ll R_{\text{on}}$):

$$f_{3\text{dB}} = \frac{1}{2\pi\, \text{I}R_{\text{on}}C_T} \tag{5-7}$$

For 5-15c ($R_{\text{on}} \ll R_S$):

$$f_{3\text{dB}} = \frac{1}{2\pi R_S(C_I + C_T)} \tag{5-8}$$

5-2.3 Source Loading Errors

Source loading errors are caused by the potentiometric effect of the source resistance and "on" resistance, with the load resistance presented by whatever follows the multiplexer. The equivalent circuit can be found in Figure 5-16.

The error expression is:

$$\%\ \text{error} = \frac{R_S + R_{\text{on}}}{R_S + R_{\text{on}} + R_L} \times 100\% \qquad (5\text{–}9)$$

For systems with similar input sources and reasonably matched switches, the problem reduces to the deviations off the nominal value of $R_S + R_{\text{on}}$. The nominal may be compensated by a simple gain change. As the "scatter" of these values becomes wider, the matching problem becomes more difficult. An example of a source loading problem is:

$$R_S = 500\ \Omega \qquad R_L = 5\ \text{M}\Omega \qquad R_{\text{on}} = 25\ \Omega$$

$$\%\ \text{error} = \frac{500 + 25}{500 + 25 + 5 \times 10^6} \times 100\% = 0.0105\%$$

In this case, the gain error for the channel due to source loading is in excess of 0.01 percent.

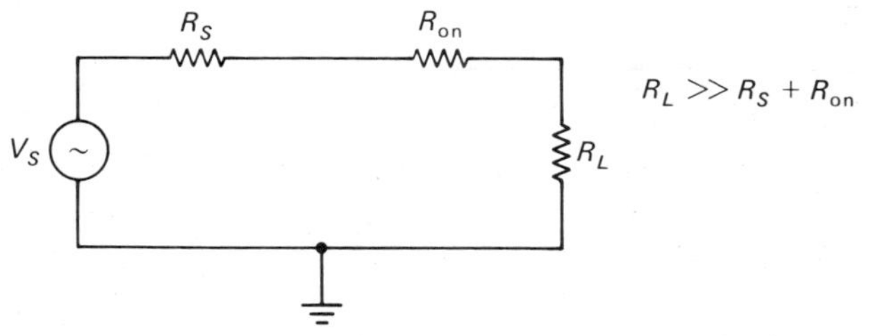

Figure 5-16. Equivalent Circuit for Source Loading

5-2.4 Crosstalk

Crosstalk is caused by signal voltages from "off" channels feeding into the measuring point through the interelement capacitances of the "off" switches. The equivalent crosstalk circuits are found in Figure 5-17. Although particular parametric values can reduce the circuit of 5-17a to those of 5-17b and 5-17c, they are still a bit more complicated than a single time constant. The problem is further complicated by the fact that for any given "on" channel, the crosstalk is a composite of feed-through from all "off" channels. If crosstalk is deemed a problem, then each channel contribution must be considered independently and the results added by superposition. This problem is nearly nonexistent in electromechanical multiplexers.

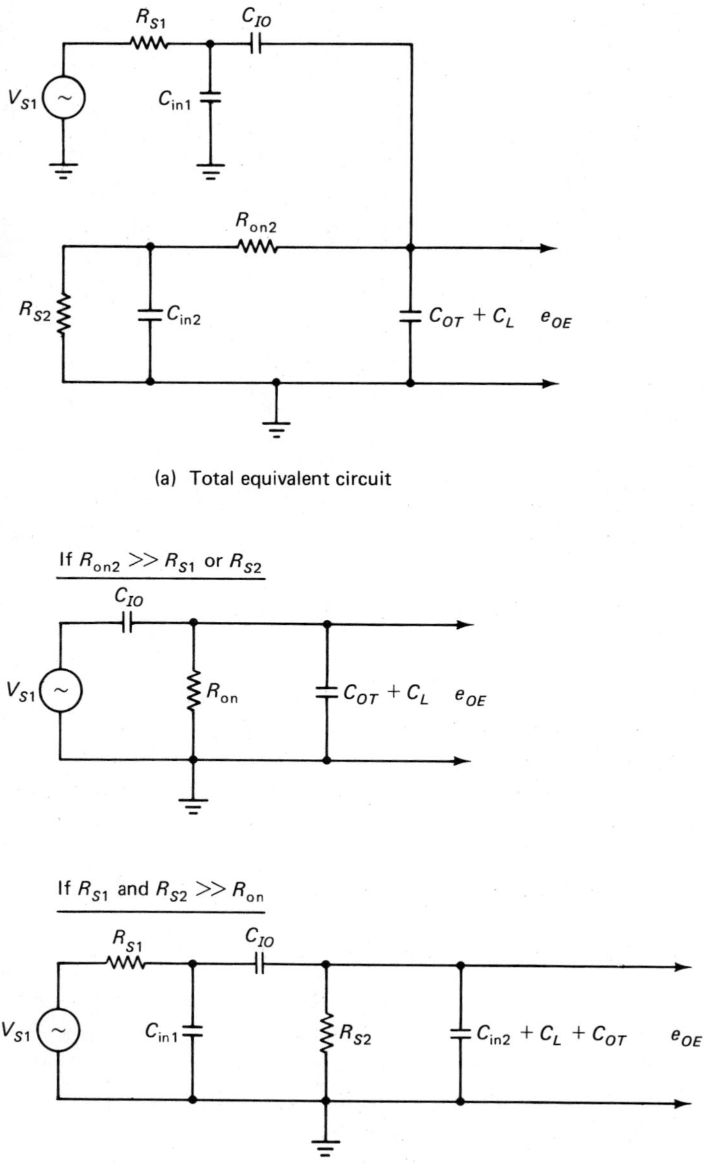

Figure 5-17. Equivalent Circuits for Crosstalk

5-2.5 Other Characteristics and Problems

You should consider other items when designing, procuring, or utilizing an analog multiplexer. These factors are either not in the manufacturer's data or very difficult to find. Some points of interest are:

1. *Understand the "on" resistance values given.*

   Often, the worst-case value is not the one provided in the specification table. Be certain that you understand variation with temperature. Also, what is the variation of R_{on} from switch to switch?

2. *Does the switch have a dead band?*

   Some CMOS devices have been known to have a point in the center of the input range where transmitting stops (the switch opens) if the gate voltage is not of a high enough value.

3. *Does the switch selected ensure break-before-make action?*

   If make-before-break should occur, the channel coming on will be momentarily connected to the channel turning off. If the overlap is sufficient, damage can occur in either of the two sources or in the switches themselves.

4. *What is power dissipation versus multiplexer speed?*

   In many semiconductor multiplexers, operating power is a function of speed. Make sure power requirements are assessed at the frequency of operation.

5. *What is the inherent thermal or inverse-frequency (I/F) noise?*

   Know what thermal or inverse-frequency noise the multiplexer itself causes, since this noise limits the lowest level signal that can be successfully measured.

5-3 MULTIPLEXER CONFIGURATIONS

Before leaving this chapter, let's consider some of the ways analog multiplexers are configured. To start off, let's look at configurations that utilize different numbers of switches per channel.

5-3.1 Multiple Switch Multiplexers

Figure 5-18 shows the simplest of implementations, in which one switch per channel is utilized. This configuration is quite acceptable when all the signals to be measured, as well as the measuring system itself, are contained within a very low-noise, laboratory-like environment. Under these conditions, the ground noise (V_E) shown in the

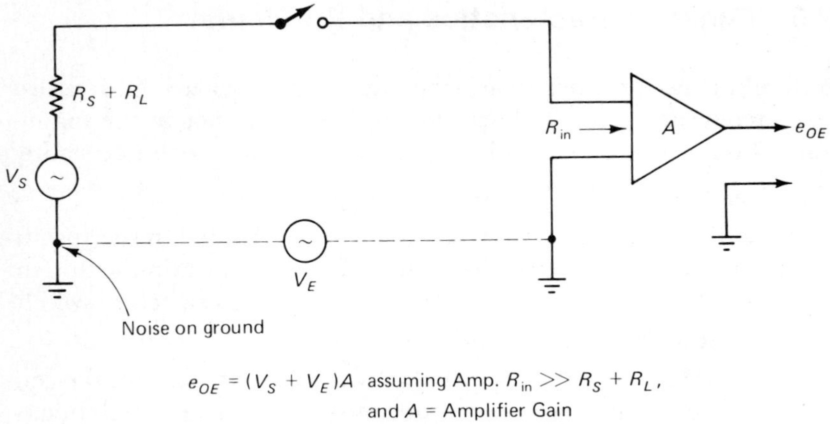

Figure 5-18. Single-Line Multiplexer Configuration

figure is negligible. Should either of the conditions be violated, we see that the amplifier output is:

$$e_{OE} = (V_S + V_E)A \tag{5-10}$$

This can be most disconcerting, since we really want to measure V_S and not the noisy environment.

Much can be gained by utilizing the balanced two-switch differential circuit shown in Figure 5-19. The noise error (V_E) appears as the same value signal at both inputs 1 and 2. The results, of course, are zero at the amplifier output, and we have successfully removed the noise effects. Unfortunately, such a perfectly balanced circuit is difficult to achieve in actual practice. To emphasize the problem, let's consider a differential, two-switch circuit with a complete unbalance (see Figure 5-20). As we can see, we now have created a real common-

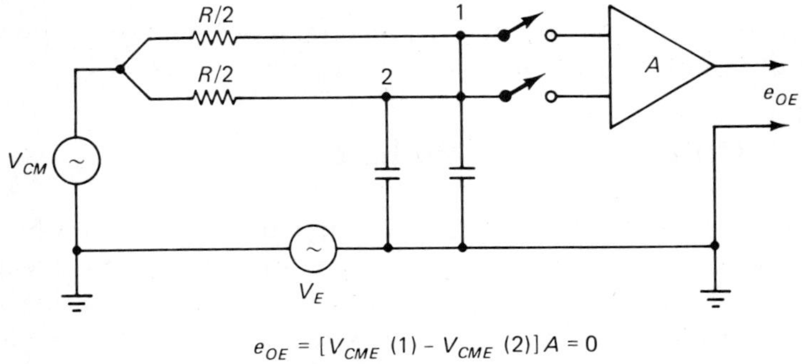

Figure 5-19. Balanced Differential (Two-Switch) Circuit

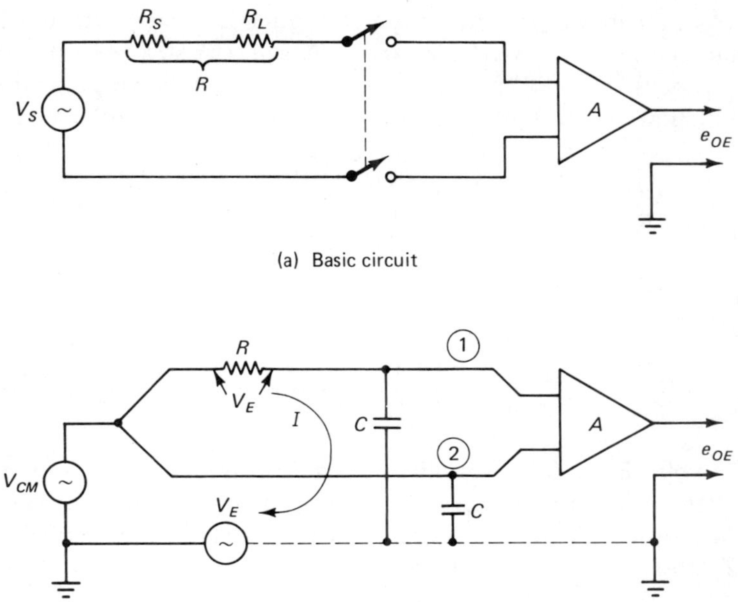

V_E adds to V_{CM}, C = cap at input to ground

(b) Equivalent common-mode circuit

Figure 5-20. Unbalanced Differential Circuit (No Shield)

mode problem. V_{CM} and V_E are totally impressed at input 2, while at input 1 their effect is reduced by the dividing action of R and C. An analysis is as follows:

Given that $X_C \gg R$, then:

$$i = \frac{V_{CM} + V_E}{X_C} = 2\pi fC(V_{CM} + V_E)$$

$$V_E = iR = 2\pi fC(V_{CM} + V_E)R$$

and the common-mode rejection ratio:

$$\text{CMRR} = \frac{V_{CM}}{V_{CME}} = \frac{V_{CM}}{V_E}$$

therefore:

$$\text{CMRR} = \frac{V_{CM}}{2\pi fCR(V_{CM} + V_E} = \frac{1}{2\pi fCR}\left[\frac{V_{CM}}{V_{CM} + V_E}\right] \qquad (5\text{–}11)$$

As we can see, the circuit performance is greatly affected by the off-balanced circuit drive resistance, R, and the stray switch and line capacitances. Using expression (5–11) and the following parameters, let's calculate the value of the stray capacitance allowable:

$$f = 60 \text{ Hz} \qquad \text{CMRR} = 10^5 \qquad V_E = 2 \text{ V}$$
$$R = 350\ \Omega \qquad V_{CM} = 10 \text{ V}$$

then:

$$C_{\max} = \frac{1}{(6.28)(6 \times 10)(3.5 \times 10^2)(10^5)} \times \frac{10}{12} \cong 63 \text{ pF}$$

at 25 to 50 pF per foot, very little hook-up wire could be used.

Let's now go one step further. Add a shield and a third switch; then multiplex the shield, as seen in Figure 5–21. This effectively shorts out V_E and equation (5–11) becomes:

$$\text{CMRR} = \frac{1}{2\pi fCR} \qquad (5\text{–}12)$$

If we use the same parameters, and recompute $C_{\max}$, we find:

$$C_{\max} = \frac{1}{(6.28)(6 \times 10)(3.5 \times 10^2)(10^5)} \cong 76 \text{ pF}$$

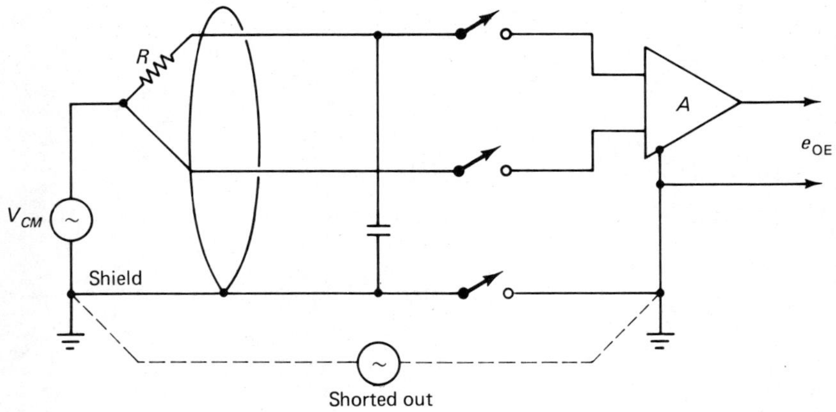

Figure 5–21. Differential Circuit with Switched Shield (Tied to Ground)

This is some improvement but still not enough. If we carry things another step further and drive the shield from the common-mode point of the input circuit, as shown in Figure 5-22, we get some dramatic results. To clearly see what happens, refer to Figure 5-23, in which we find the same circuit with a complete drive unbalance. The stray capacitances of the lines and the switches are referenced to the shield which, right at the multiplexer output, now differs by a small amount, $(1 - \alpha)$, from V_{CM}.

This being the case, the following derivation can be made:

With $X_C \gg R$:

$$i = \frac{(1 - \alpha)V_{CM}}{X_C} \qquad \text{and} \qquad (1 - \alpha) \ll 1$$

so that:

$$\text{CMRR} = \frac{1}{2\pi f C(1 - \alpha)R} \tag{5-13}$$

Using the same parameters as the previous example and considering that $(1 - \alpha)$ for good cable is 0.001 and $C = 50{,}000$ pF for 1,000 feet of cable,

$$\text{CMRR} = \frac{1}{(6.28)(6 \times 10)(5 \times 10^{-8})(10^{-3})(3.5 \times 10^{2})} = 1.54 \times 10^{5}$$

This is a substantial improvement. The CMRR required by the previous example has been exceeded by 50 percent—with 1,000 feet of cable. With this technique, systems with 120 dB of common-mode rejection ($\text{CMRR} = 10^{6}$) have been developed.

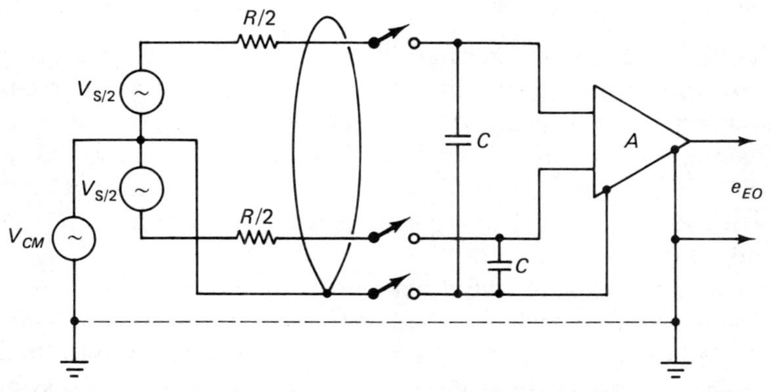

Figure 5-22. Differential Circuit with Switched Shield Driven

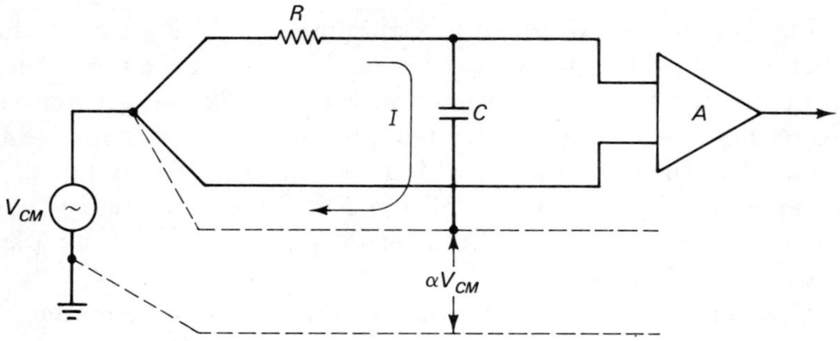

Figure 5-23. Equivalent Common-Mode Circuit with Unbalanced, Differential-Guarded Connection

5-3.2 The Flying Capacitor Multiplexer

Another technique for achieving high common-mode rejection is the flying capacitor switch configuration shown in Figure 5-24. This switch is a very slow-speed, low-level multiplexer that uses reed relays. Under normal conditions, each channel capacitor (C_1) is attached to its associated input, where it is following the signal to be measured. When a channel is to be converted, C_1 is switched from its normal position to the common measurement bus. The sequence of events is:

1. The previous channel capacitor is switched off the common bus.
2. Then relay RY_3 is open and RY_4 closed. This removes any residual charge left on C_2 by the previous channel.
3. While this is going on, the new channel is switched onto the common point by its channel relays (RY_1 and RY_2).
4. To finish off the sequence, RY_4 is opened and RY_3 closed to connect the measurement point to the measuring system.

As the figure shows, the normal circuit (the switch connected to the source) forms a bridge: the better the balance, the less the common-mode interference affects the measurement. C_1 is a very large capacitor, having values in the order of 100 μF. C_2, which is used to swamp out the variations and amplifier input capacitance, is usually in the order of 100 pF. The circuit is generally slow, providing sample rates of around 50 channels/second.

It also comes with one other undesirable speed parameter: It is limited to about 10 samples/channel/second. This slowness is caused by the need to charge C_2 from C_1 each time it is switched into the common point, as well as the need to recharge C_1 each time it is switched back to its normal position. Common-mode rejection of 120 dB can be

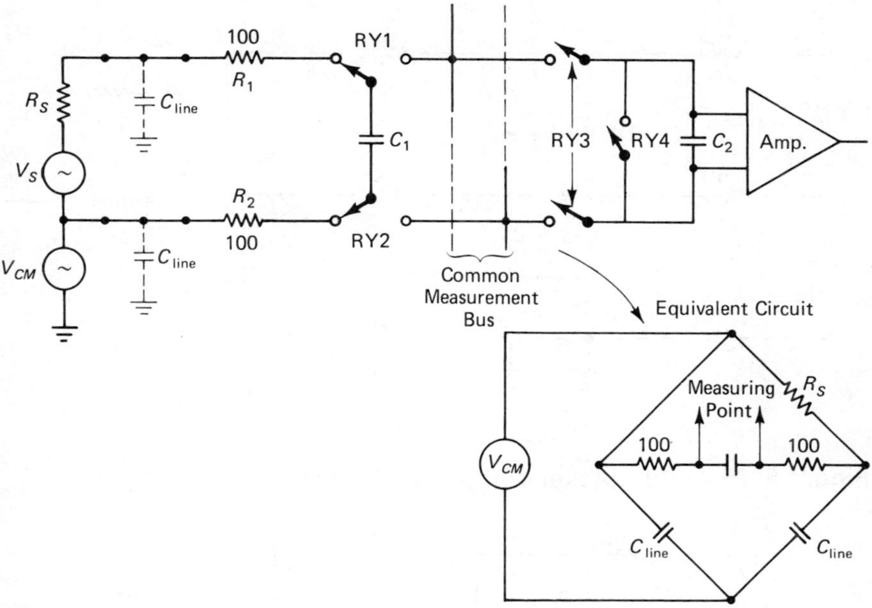

Figure 5-24. The Flying Capacitor Multiplexer

achieved with reasonably long lines and input unbalance approaching 1,000 Ω. The circuit is extremely good for monitoring dc or slow varying dc, under very severe noise conditions. The small resistors, R_1 and R_2, are to prevent the large capacitor, C_1, from shorting out the source on turn-on.

5-3.3 Other Configurations

Figures 5-25 and 5-26 show two other multiplexer configuration alternatives. Both circuits work directly into the summation point of an inverting operational amplifier (Chapter 6). The diodes in Figure 5-25 maintain the left-hand side of the switch within +0.5 V of ground at all times.

Since the summation point is also close to ground, the switch never sees a significant voltage across it, thus substantially reducing the settling time problem. Even the switching transients are looking into a low impedance point, if the amplifier has significant bandwidth. The circuit of Figure 5-26 is simply a variation of that in Figure 5-25. The only difference is that the clamp is provided by a switched transistor rather than by diodes. The advantage is that the clamp action is more positive—the diodes have a transition point at which neither is really on.

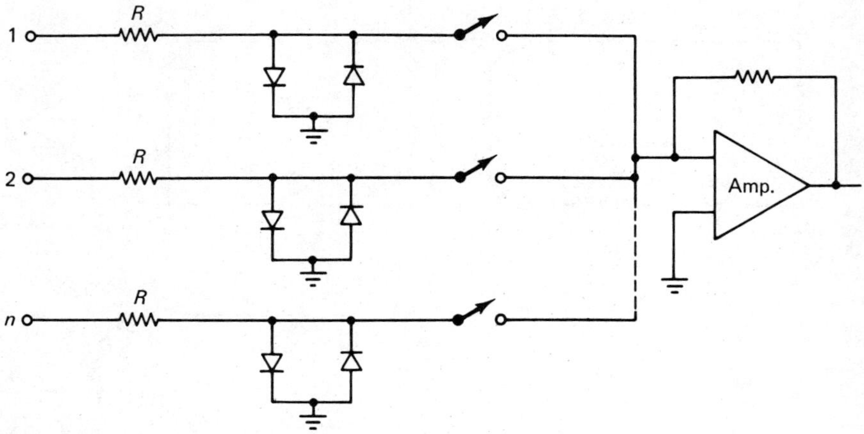

Figure 5-25. Multiplexer with Diode Clamps

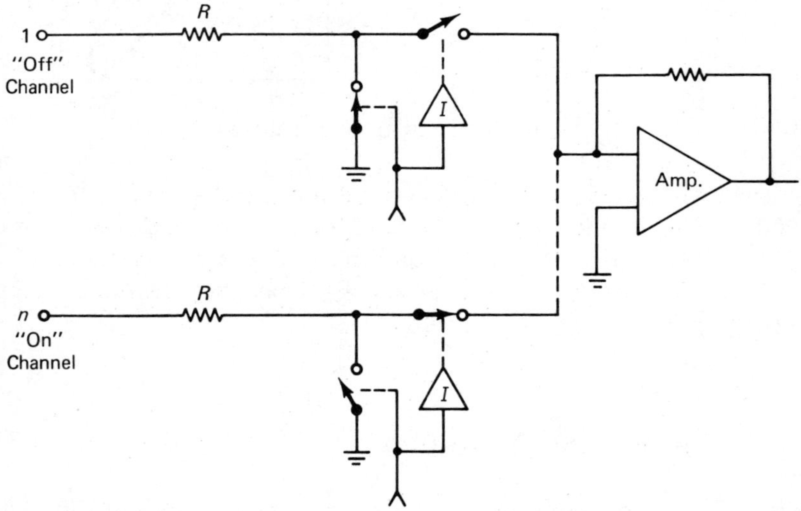

Figure 5-26. Multiplexer with Transistor Clamps

PROBLEMS

5-1. A data acquisition system has 10 channels of analog multiplexing. Each channel contains frequencies up to 5 kHz and must be sampled at a rate of at least 10 samples/cycle.

a. What is the multiplex sample rate?

b. Which multiplexer switch types would be candidates for this application?

5–2. A data acquisition system requires 20 channels of analog multiplexing. Signal levels are in the 100-mV range and high accuracy requirements dictate switch-on resistance in the sub-Ω range. Each channel contains frequencies up to 2.5 Hz and requires sampling of at least 5 samples/cycle.
a. What is the composite multiplex rate?
b. What would be a candidate switch type?

5–3. A 16-channel analog multiplexer has the following characteristics:

$R_{on} = 20\ \Omega$	$C_L = 50$ pF
$C_I = 10$ pF	$R_L = 10$ MΩ
$C_O = 10$ pF, each switch	$I_L = 1$ nA @ 25° C

a. With the source resistance, R_S, of the "on" channel = 1 kΩ, what is the dc error due to source loading and leakage current?
b. What is the settling time to 0.1 percent (no turn-on transient)?
c. What is the ac response of the "on" switch?

5–4. If the multiplexer of problem 5–3 were constructed with four tiers of four switches, recalculate the leakage error and the settling time. (*Note*: R_{on} of second-tier switches is also 20 Ω.)

5–5. Derive expression (5–5).

5–6. A data acquisition system uses a single-line multiplexer as shown in Figure 5–18. The full-scale input voltage is + 10 V, and the system converts data to 10 bits binary. The amplifier gain is 10. A given channel has ground noise (V_E) of 5 mV RMS. Can this system successfully measure the designated channel input to the resolution implied by the number of bits?

5–7. A multiplexer uses two switches per channel, as seen in Figure 5–20, and is driven by a source unbalance of 1,000 Ω. The noise error sources are $V_{CM} = 5$ V and $V_E = 1$ V, both at 60 Hz. (*Note*: Assume cable capacitance is 25 pF/ft in all cases.)
a. If the hook-up cable has a capacitance of 25 pF/ft, and if the required CMRR is 10^4, how far can the transducer be from the data acquisition system?
b. If a shield were used, as shown in Figure 5–21, what would be its length?
c. What if the shield were switched, as shown in Figure 5–22, and the value of $(1 - \alpha)$ were 0.01?

REFERENCES

1. Siliconix, Inc., *Analog Switches and Their Applications*. Santa Clara, Cal.: Siliconix, Inc., 1976.
2. P. H. Barrett, *Analog Systems for Microprocessors and Minicomputers*. Reston, Va.: Reston Publishing Company, 1978.
3. D. F. Hoeschele, *Analog-to-Digital/Digital-to-Analog Conversion Techniques*. New York: John Wiley & Son, Inc., 1968.
4. H. Schmid, *Electronic Analog/Digital Conversions*. New York: Van Nostrand Reinhold Company, 1970.
5. D. H. Sheingold, *Analog–Digital Conversion Handbook*. Norwood, Mass.: Analog Devices, Inc., 1972.

6
Signal Conditioning

6-0 INTRODUCTION

Almost any signal we encounter requires some amount of conditioning before it is presented for analog-to-digital conversion. The conditioning can be as simple as buffering to provide an impedance translation between the signal source and the conversion point. Or it can be as complicated as a very sophisticated sample-and-hold, where infinitely narrow samples are taken at precise intervals.

Like the analog multiplexer discussed in Chapter 5, the signal conditioner operates in the analog domain and is therefore subject to all the limitations of the analog world. Care must be taken that the undesirable effects do not outweigh the benefits of the function being used.

This chapter considers the signal conditioners used in the typical data acquisition system. The first part of the chapter is devoted to the various types of amplifier configurations; the latter part covers the sample-and-hold and the peak-detect-and-hold functions. You will be provided with enough background to be able to specify, select, procure, and use these devices.

6-1 THE INVERTING OPERATIONAL AMPLIFIER

Perhaps the most well known of the available signal conditioning devices is the operational amplifier. Since this workhorse was the mainstay of the analog computer industry for years, it naturally became a building block as attention turned toward preparing data for digital computing. The prime characteristics of the classical operational amplifier (op-amp) are:

1. very high input impedance (typically 10^7 to $10^{12}\ \Omega$),
2. low output impedance (<1 to $100\ \Omega$),
3. very high gain (100 K to 1 MEG),
4. an open-loop response that rolls off to zero, after the 3 dB point, at a 6 dB/octave rate.

The only present-day deviation from these characteristics comes in the area of open-loop response, where the shape and roll-off have often been tailored to enhance slew rate and settling time. You must exercise care when using such devices, to be sure they remain stable under closed-loop conditions.

The operational amplifier may be used in both inverting and noninverting modes. In this section we will discuss the inverting amplifier, while the next section will cover the noninverting configuration.

6-1.1 Gain

Figure 6-1 shows the operational amplifier connected as an inverting device. Summing the currents shown, we have:

$$\frac{V_S - V_E}{R_I} + \frac{V_O - V_E}{R_F} - I_E = 0 \qquad (6\text{–}1)$$

If the amplifier open-loop gain is very high, and I_E is negligibly small, the two amplifier inputs will approach the same voltage value, and a *virtual ground* is created. V_E, then, also approaches zero. Under these conditions, expression (6-1) reduces to:

$$\frac{V_S}{R_I} + \frac{V_O}{R_F} = 0 \qquad (6\text{–}2)$$

and

$$A_{CL} = \frac{V_O}{V_S} = -\frac{R_F}{R_I} \qquad (6\text{–}3)$$

where A_{CL} = closed loop gain

The exact expression for the closed-loop gain is:

$$A_{CL} = -\frac{R_F}{R_I}\left[\frac{1}{1 + \frac{R_I + R_F}{R_I A_{OL}}}\right] \qquad (6\text{–}4)$$

where A_{OL} = the amplifier open-loop gain

$\frac{R_I A_{OL}}{R_I + R_F}$ = the gain without the closed loop (loop gain)

Obviously, the open-loop gain value has a significant impact on the validity of expression (6–3). It also has a great impact on the *virtual ground* created at the amplifier input terminals. Given the conditions of a good op-amp, then $I_1 = I_2$ and a small change in V_E will cause a large change in I_2 as shown:

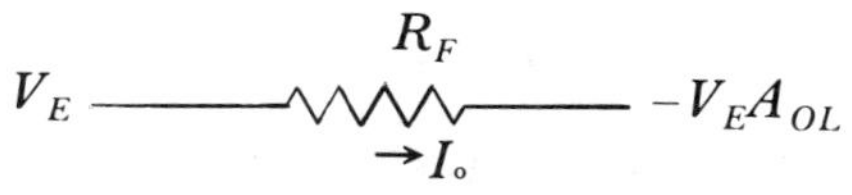

where
$$I_2 = \frac{V_E + V_E A_{OL}}{R_F} = \frac{V_E(1 + A_{OL})}{R_F} \qquad (6\text{–}5)$$

$$R_{inCL} = \frac{V_E}{I_2} = \frac{R_F}{1 + A_{OL}} \qquad (6\text{–}6)$$

So we see that, as A_{OL} gets higher, R_{inCL} looks more like zero, and thus a virtual ground.

Looking back at expression (6–3), we see that for a good operational amplifier, the closed-loop gain is totally dependent on the input and feedback resistors. A common mistake is to forget that the amplifier may be driven by a source with a finite output resistance. This could occur if an analog multiplexer were connected between the signal source and the amplifier's input terminals. Figure 6–2 shows such a situation, where the input signal source, V_S, is connected to point 1 through the resistance, R_S. So the proper closed-loop gain expression is:

$$A_{CL} = -\frac{R_F}{R_S + R_I} \qquad (6\text{–}7)$$

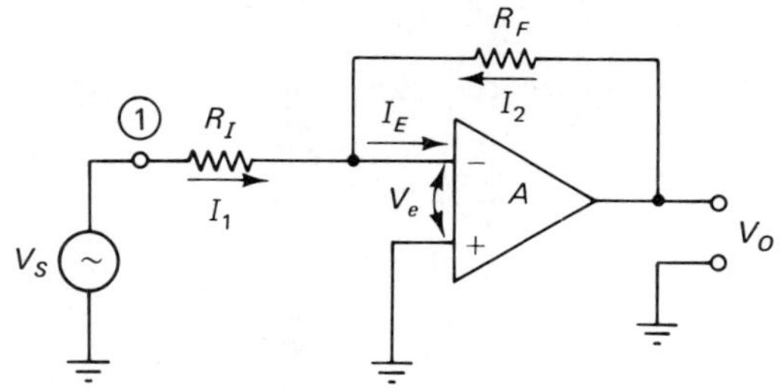

Figure 6-1. Inverting Operational Amplifier

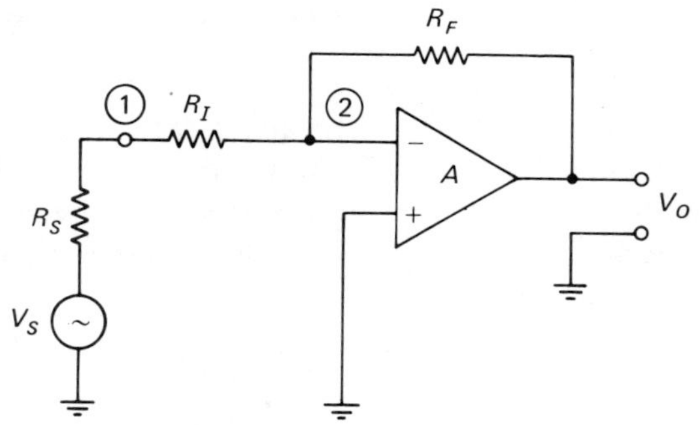

Figure 6-2. Inverting Op-Amp with Source Resistance

As the amplifier (Figure 6-2) has a virtual ground at point 2, the effective input resistance seen by the source is the resistor, R_I. The output resistance is the basic amplifier output resistance divided by the loop gain, or:

$$R_{O\text{-}CL} = \frac{R_{O\text{-}CL}}{\text{Loop gain}} \qquad (6\text{–}8)$$

6-1.2 Offset Voltage

Another parameter of concern is *offset voltage*. As Figure 6-3 shows, the op-amp has a finite voltage offset, V_{OS}, at its terminals, which represents the mismatch between the two input terminals. In the worst case for bipolar units, this is a few tens of microvolts, while for good field-effect devices, it can be as low as a few tens of millivolts. For room temperature compensation, the combination of R_a and R_b are added. As point 1 is a virtual ground, the resistor combination does not affect signal path performance. Although compensation can be effected at room temperature, V_{OS} changes with temperature. Ignoring the effects of R_a and R_b, the effect of ΔV_{OS} is:

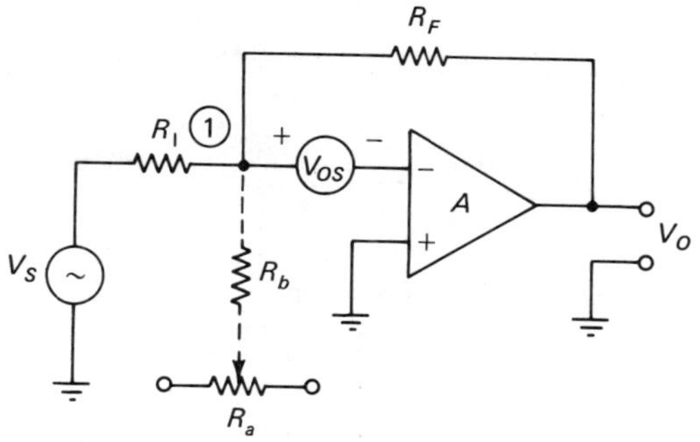

Figure 6-3. Inverting with Offset Adjustment

RTO (referred to output):

$$\Delta V_O = \Delta V_{OS}\left[\frac{R_I + R_F}{R_I}\right] \tag{6–9}$$

or:

$$\Delta V_O = \Delta V_{OS} \quad 1 + A_{CL} \tag{6–10}$$

RTI (referred to input):

$$\Delta e_I = -\Delta V_{OS}\left[\frac{R_I + R_F}{R_F}\right] \tag{6–11}$$

6-1.3 Bias and Offset Current

Bias current can be seen exiting each of the input terminals of the amplifier shown in Figure 6-4. Consider the condition where R_C = zero. I_b, entering node 1, would create an offset voltage:

$$V_{OS\text{-}I} = I_{b1}\left[\frac{R_I R_F}{R_I + R_F}\right] \tag{6–12}$$

Substituting into expression (6–9) and solving for V_O, rather than ΔV_O, we have:

$$V_{O\text{-}I_b} = I_{b1} R_F \tag{6–13}$$

As R_F becomes larger, the condition becomes worse. Suppose, however, we allowed R_C to have a value. The optimum situation would occur when:

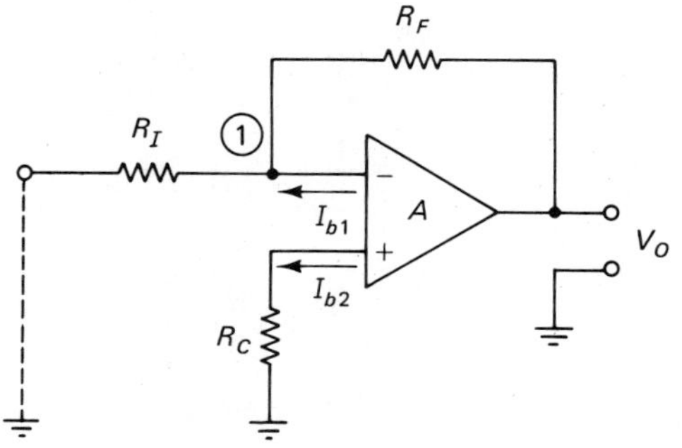

Figure 6-4. Inverting Amplifier with Drift Compensation

$$V_{O\text{-}I_{b1}} = V_{O\text{-}I_{b2}}$$

If $I_{b1} = I_{b2}$, this will occur when:

$$R_C = \frac{R_F R_I}{R_F + R_I} \qquad (6\text{–}14)$$

Now only the difference between the two bias currents—that is, I_{OS} or the offset current—is a problem. For this situation, I_{OS} is substituted into (6–13) to form:

$$V_{O\text{-}I} = I_{OS} R_F \qquad (6\text{–}15)$$

This is advantageous because the offset current is usually considerably smaller than the bias currents. Note that offset and bias currents increase with temperature.

6-1.4 Frequency Response

Figure 6-5 contains plots of both the open- and closed-loop frequency responses for a single-pole op-amp that has been connected as shown in Figure 6-1. The closed-loop gain has been selected to be somewhat less than the open-loop gain. As we see, the closed-loop response remains flat, as the frequency increases, until the loop gain becomes less than one. At that point, the closed-loop response starts to follow the open-loop response, and it too rolls off. As you can see, with a sacrifice in gain, the closed-loop bandwidth has been extended well beyond that of the open-loop response. This trade-off works well, providing that the open-loop roll-off is 6 dB/octave, thus insuring no greater than 90° of additional phase shift. If the open-loop response had contained more

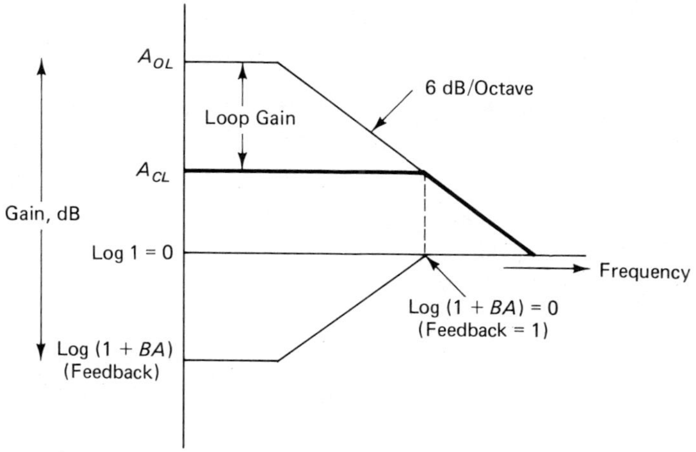

Figure 6–5. Gain Versus Frequency Plot for Single-Pole Amplifier

than one real pole and the additional phase shift had exceeded 120° at the loop gain plot zero-crossing, a decided peak would appear in the closed-loop response. In other words, the loop gain must be less than unity before the excess-phase-shift exceeds 120°.

The curves of Figure 6–5 imply small signal operation. When large signals are required, the bandwidth shrinks in proportion to the output voltage amplitude. This problem, which is related to the internal parametric capacitances of the amplifier, also has an impact on the amplifier's ability to slew from minus full-scale output to plus full-scale. Known as "slew rate," this parameter is defined by:

$$T_{SR} = \frac{\Delta V}{\Delta T} = 2\pi f_p \, V_{PP} \qquad (6\text{–}16)$$

where T_{SR} = slew rate

$f_{3\,\mathrm{dB}}$ = the maximum frequency at unity closed-loop gain for the given peak-to-peak output, with acceptable distortion (usually less than 3%)

V_{PP} = peak-to-peak voltage

The amplifier's ability to settle to some value within a given accuracy (Figure 6–6) is of great interest to the user of a data acquisition system, for it determines when measurement may successfully begin. The definition of settling time is:

> the time elapsed from the application of a full-scale step input to when the output has settled to within a specified error band around its final value.

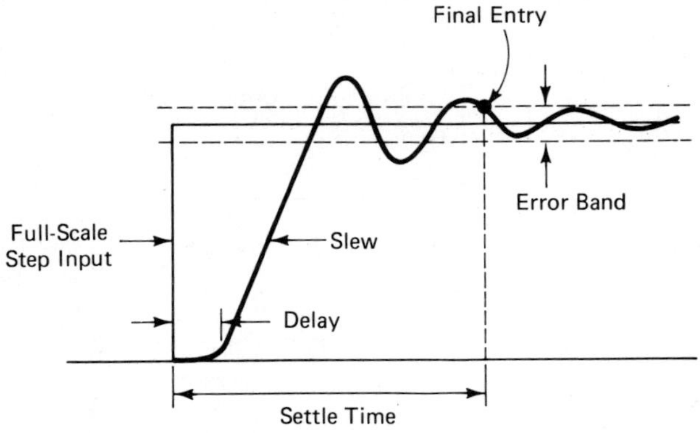

Figure 6-6. Amplifier Settling Time

Settling time is comprised of three parts:

1. a delay that occurs before the output starts moving,
2. the slew time, and
3. the time to final entry into the allowable error zone.

The settling time is always at least as long as that depicted by its full-power frequency. Consider the following example, in which you are given:

Maximum full-power frequency:	6 MHz
Slew rate:	140 V/μsec
Full-voltage swing:	±10 V (20 V pp)
Start-up delay:	35 nsec

What is the settling time to 0.01 percent?

1. First calculate the basic system time constant:

$$TC = \frac{1}{2\pi \text{Bw}} = \frac{1}{(6.28)(6 \times 10^6)} = 26.5 \text{ nsec}$$

2. Settle to 0.01 percent requires 9.2 TC or:

$$\text{Total } TC = 26.5 \times 9.2 = 244.0 \text{ nsec}$$

3. Calculate the time due to the slew rate:

$$T_{\text{slew}} = \frac{\Delta V}{\text{Slew rate}} = \frac{20 \text{ V}}{140 \text{ V}/\mu\text{sec}} = 143 \text{ nsec}$$

4. The total settling time is:

$$T_{\text{settle}} = TC + T_{\text{slew}} + T_{\text{start-up}}$$
$$= 244.0 + 143 + 35 \cong 422 \text{ nsec}$$

One particularly annoying problem with the inverting op-amp is its susceptibility to stray capacitance, which appears across the feedback resistor. As Figure 6-7 shows, a small amount of stray capacitance can cause the closed-loop response to roll down much earlier than the open-loop response would imply. To understand the significance of this, consider the following example:

$R_I = 1\ \text{M}\Omega$ (not high for a typical DAS application)

$A_{CL} = 10$ A

$C_{\text{stray}} = 1$ pF

$$R_F = A_{CL}R_I \qquad (6\text{–}17a)$$
$$= (10)(10^6) = 10^7\ \Omega$$

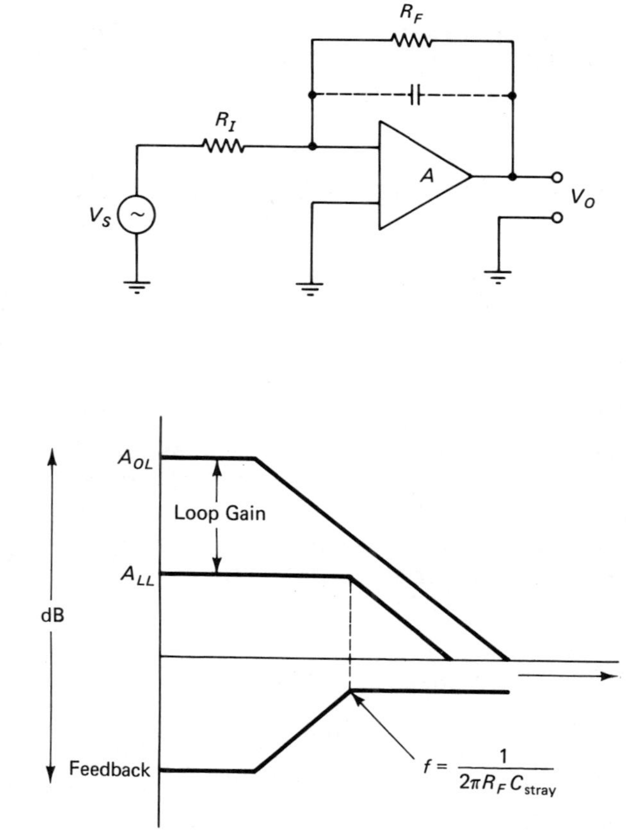

Figure 6-7. Effect of Stray Capacitance on the Inverting Amplifier

$$f_{3\,\text{dB}} = \frac{1}{2\pi R_F C_{\text{stray}}} \qquad (6\text{–}17b)$$

$$= \frac{1}{(6.28)(10^7)(10^{-12})}\ 16\ \text{kHz}$$

As we see, a small amount of stray capacitance can severely reduce the closed-loop bandwidth of an amplifier when high resistances are used. One of the advantages of the inverting amplifier is that it is extremely versatile where operational functions—such as signal addition, subtraction, and integration—are concerned.

6–2 THE NONINVERTING OPERATIONAL AMPLIFIER

The often used noninverting operational amplifier (op-amp) configuration can be seen in Figure 6–8. If A_{OL} and R_{in} are both high (as we would expect for a good op-amp), then V_F would equal the source voltage, V_S. If such is the case, then:

$$V_F = V_S = V_O \frac{R_2}{R_1 + R_2}$$

and

$$A_{CL} = \frac{V_O}{V_S} = \frac{R_1 + R_2}{R_2} \qquad (6\text{–}18)$$

Should R_1 go to zero (the output shorted to the negative input) as shown in Figure 6–9, the amplifier closed-loop gain would be unity.

A major feature of the noninverting amplifier is its inherent, very high input resistance. As it turns out, the input resistance is bootstrapped by the feedback and effectively amplified by the loop gain as shown in the following expression:

$$R_{\text{in-e}} = \underbrace{\frac{A_{OL}}{A_{CL}}}_{\text{Loop gain}} R_{\text{in}} \qquad (6\text{–}19)$$

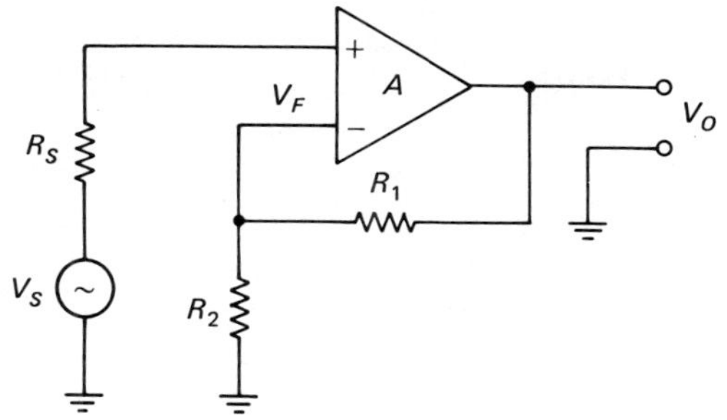

Figure 6-8. The Noninverting Operational Amplifier

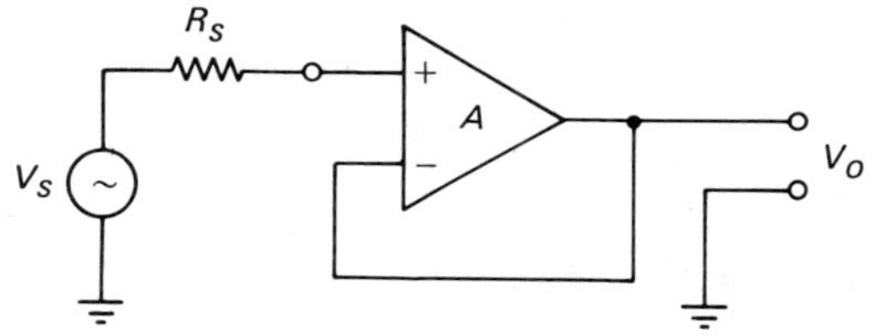

Figure 6-9. Noninverting, Unity-Gain Amplifier

where $R_{in\text{-}e}$ = the effective input resistance
R_{in} = the basic input resistance of the amplifier.

Consider the following examples, in which you are given:

$$R_{in} = 100\ \text{k}\Omega \qquad R_2 = 10\ \text{k}\Omega$$
$$A_{OL} = 10^5 \qquad R_1 = 1\ \text{k}\Omega$$

$$A_{CL} = \frac{10\ \text{k}\Omega + 1\ \text{k}\Omega}{1\ \text{k}\Omega} = 11 \qquad (a)$$

$$R_i = \frac{10^5}{11}[100\ \text{k}\Omega] \cong 900\ \text{M}\Omega \qquad (b)$$

This represents an increase of about 10^4 times. Actually, R_{in} is in parallel with the common-mode input resistance, R_{CM}, which exists between the positive input terminal and ground (Figure 6-10). If R_{CM} were 1,000 MΩ (a typical value), the effective input resistance would still be in excess of 450 MΩ.

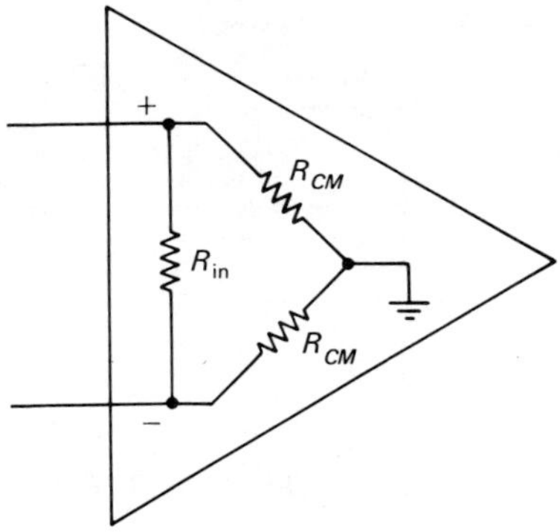

Figure 6-10. Amplifier Input

6-2.1 Frequency Response

Stray capacitance across the feedback resistor (R_1) must also be considered for the noninverting configuration. Consider first a 1 pF capacitance. Typical values for a gain of 10 might be $R_1 = 10\ \text{k}\Omega$ and $R_2 = 900\ \Omega$. The bandwidth restriction brought about by the R_1/C_{stray} combination would be:

$$f_{3\text{ dB}} = \frac{1}{2\pi R_1 C_{\text{stray}}} = \frac{1}{(6.28)(10^4)(10^{-12})} = 16\text{ MHz}$$

This has considerably less effect than a similarly sized capacitor across the feedback resistor in an inverting op-amp. This feature of the noninverting op-amp allows high gains to be achieved with small resistance values, while still maintaining a high input resistance.

Stray capacitance across R_2 should be carefully observed, however, since it adds an extra pole in the feedback loop. The effect is an increase in feedback phase shift and a resultant peak in the closed-loop response, as seen in Figure 6-11. Fortunately, R_2 can usually be made small in value, and the unwanted pole can be moved to a point of little significance.

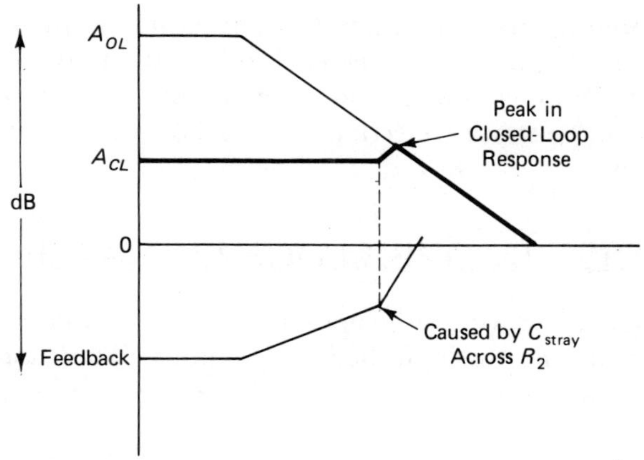

Figure 6-11. Effect of Stray Capacitance Across R_2 in the Noninverting Amplifier

6-2.2 Offset Voltage and Offset Current

Like its inverting counterpart, the noninverting op-amp is susceptible to offset voltages and offset currents. The expressions for these factors are as follows:

Offset Voltage:

RTI (referred to input)

V_{OS}

RTO (referred to output)

$$\Delta V_O = \Delta V_{OS} \left[\frac{R_2 + R_1}{R_2}\right] \tag{6–20}$$

Offset Current:

RTI

$$\Delta V_{I\text{-}OS} = I_{b1}R_S - I_{b2}\left[\frac{R_1R_2}{R_1 + R_2}\right] \tag{6–21}$$

RTO

$$\Delta V_{O\text{-}OS} = \Delta V_{I\text{-}OS}\left[\frac{R_2 + R_1}{R_2}\right] \tag{6–22}$$

The noninverting amplifier has one other peculiarity. Because the two inputs must always track each other, the normal mode signal also appears as a common-mode signal. An op-amp with a good internal common-mode rejection ratio (CMRR) is a must for this configuration.

6-3 NOISE IN OPERATIONAL AMPLIFIERS

Any noise generated within the operational amplifier limits the smallest signal that may be distinguished at its input. The basic amplifier, along with its associated noise generator, can be seen in Figure 6-12a.

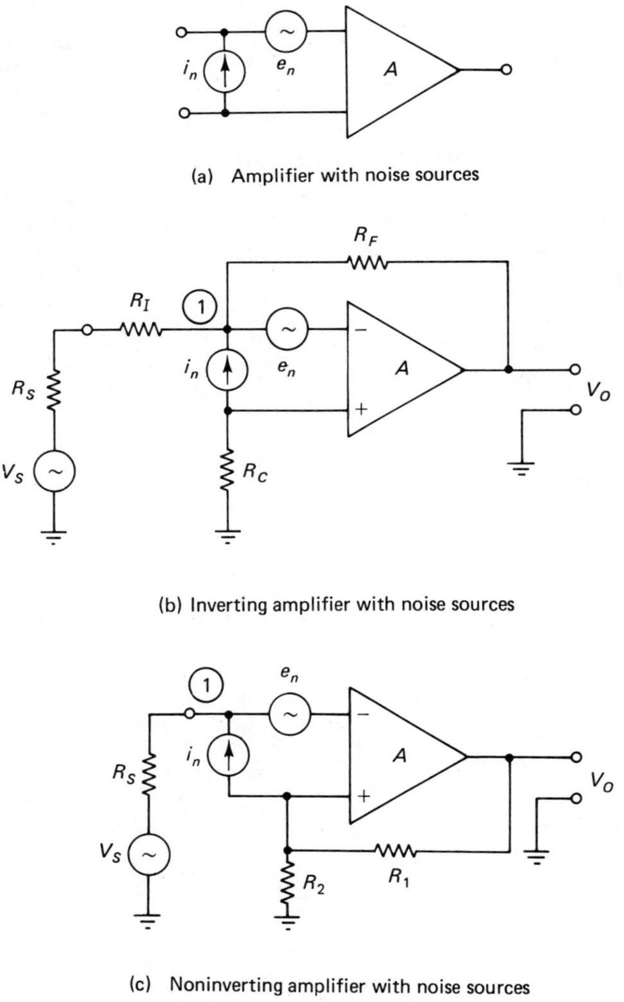

Figure 6-12. Operational Amplifiers with Noise Sources

The voltage noise source predominates when low source impedances are used, while the current source is the major factor for high R_S. Although these sources are not totally independent, their correlation is quite small.

6-3.1 The Inverting Amplifier

Figure 6-12b contains the inverting op-amp with the noise sources identified. The noise voltage referred to the amplifier input, point 1, is:

$$e_{nI_T} = \sqrt{4KTR_T\Delta f} + I_n R_T + e_n \qquad (6\text{–}23)$$

where $\sqrt{4KTR_T\Delta f}$ = the resistor noise

$$R_T = \frac{(R_I + R_S)R_F}{R_I + R_S + R_F} + R_C$$

K = Boltzmann's constant = 1.38×10^{-23} joules

T = temperature in degrees Kelvin (° K)

Δf = the amplifier bandwidth

At the output of the amplifier, the noise voltage is:

$$E_{nO_T} = e_{nI_T} \frac{R_I + R_S + R_F}{R_I + R_S} \qquad (6\text{–}24)$$

Let's consider the following example, in which you are given:

R_I = 95 kΩ	e_n = 1.5 μV (RMS)
R_S = 5 kΩ	i_n = 10 pA (RMS)
R_F = 100 kΩ	T = 300° K (25° C)
Δ_f = 10 kHz	R_C = 50 kΩ

Note that A_{CL} is 1.

$$\begin{aligned} e_{nT} &= \sqrt{4(1.38 \times 10^{-23})(3 \times 10^2)(10^5)(10^4)} \\ &\quad + (10^{-11})(10^5) + 1.5 \times 10^{-6} \\ &= 4.07 \times 10^{-6} + 1.0 \times 10^{-6} + 1.5 \times 10^{-6} \\ &= 6.57\ \mu V \end{aligned}$$

At the amplifier output:

$$E_{nO_T} = (6.57)\left[\frac{200\ k\Omega}{100\ k\Omega}\right] = 13.14\ \mu V$$

6-3.2 The Noninverting Amplifier

The noninverting case can be found in Figure 6-12c. The noise voltage, referred to point 1, is:

$$e_{nI_T} = \sqrt{4KTR_t\Delta f} + I_nR_t + e_n \qquad (6\text{–}25)$$

where

$$R_t = R_S + \frac{R_1R_2}{R_1 + R_2}$$

The only difference between this expression and (6-24) is that R_t has been substituted for R_T in the first two factors. The noise voltage at the amplifier output, for this case, is:

$$E_{noT} = e_{nI_T}\left[\frac{R_1 + R_2}{R_1}\right] \qquad (6\text{–}26)$$

We will consider an example that is comparable to the one for the inverting circuit. Given:

$$\begin{aligned} R_1 &= 1\ \text{k}\Omega & e_n &= 1.5\ 1.5\ \mu\text{V} \\ R_2 &= 10\ \text{k}\Omega & i_n &= 10\ \text{pA} \\ R_S &= 5\ \text{k}\Omega & T &= 300°\ \text{K}\ (25°\ \text{C}) \\ \Delta_f &= 10\ \text{kHz} \end{aligned}$$

Note that the gain is 1.1.

$$\begin{aligned} e_{nI_T} &= \sqrt{4(1.38 \times 10^{-23})(3 \times 10^2)(5.9 \times 10^3)(10^4)} \\ &\quad + (10^{-11})(5.9 \times 10^3) + 1.5 \times 10^{-6} \\ &= 0.988 \times 10^{-6} + 0.059 \times 10^{-6} + 1.5 \times 10^{-6} \\ &= 2.55\ \mu\text{V} \end{aligned}$$

At the amplifier output:

$$E_{noT} = 2.55\ \frac{1\ \text{k}\Omega + 10\ \text{k}\Omega}{10\ \text{k}\Omega} \cong 2.8\ \mu\text{V}$$

In general, compared to the inverting circuit, the noninverting circuit has superior noise characteristics, given the same gain and bandwidth.

At low frequencies, the noise is thermal, or $1/f$ noise, which starts to roll up at about 1 kHz.

6-4 COMMON-MODE REJECTION

Differential or two-input amplifiers can be considered to contain two separate paths that are combined somewhere within the amplifier (Figure 6-13). If path A_1 is not exactly the same as path A_2, an error voltage (V_{OCM}) appears at the output when the same voltage (V_{CM}) is applied to the input.

The expression for the output error voltage is:

$$V_{OCM} = V_{CM}(A_2 - A_1) \qquad (6\text{–}27)$$

The equivalent differential input voltage required to produce the same output voltage is:

$$e_{CM} = \frac{V_{OCM}}{\text{Differential gain}} = \frac{V_{OCM}}{A} \qquad (6\text{–}28)$$

The expression for the common-mode rejection ratio (CMRR) is:

$$\text{CMRR} = \frac{\text{Differential gain}}{\text{Common-mode gain}} = \frac{A}{A_2 - A_1} \qquad (6\text{–}29)$$

If the expression (6–27) is substituted into (6–28), we find:

$$e_{CM} = V_{CM}\frac{(A_2 - A_1)}{A} = \frac{V_{CM}}{\text{CMRR}} \qquad (6\text{–}30)$$

Equipped with these expressions, let's now consider an example where calculation of the output error voltage, V_{OCM}, is desired. Given:

$$V_{CM} = 10 \text{ V}$$

$$\text{CMR} = 120 \text{ dB, CMRR} = 10^6$$

$$A = 1{,}000$$

then:

$$e_{CM} = \frac{10 \text{ V}}{10^6} = 10\ \mu\text{V}$$

and:

$$V_{OCM} = (1{,}000)(10^{-5}) = 10 \text{ mV}$$

Keep in mind that this section has been limited to only the inherent common-mode problems of a basic amplifier. As we discovered in Chapter 5, improper multiplexer connections can also contribute to the common-mode problem. As we shall see in the following section, components that establish the gain and other signal conditioner operations also reduce the total system CMRR.

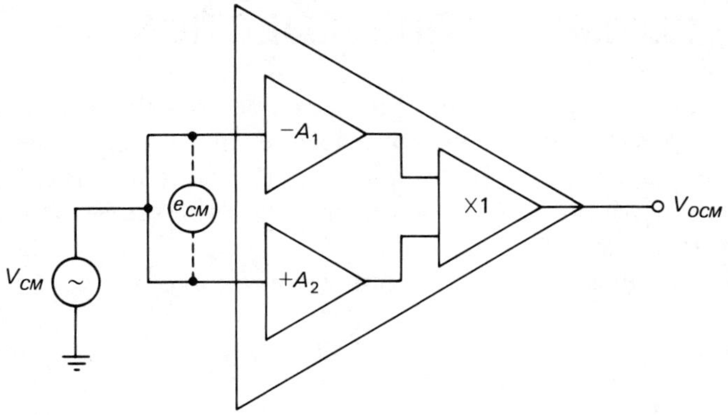

Figure 6-13. Equivalent Common-Mode Circuit

6-5 DIFFERENTIAL OPERATIONAL AMPLIFIERS

The amplifier circuits discussed up to this point have all been single-input configurations. Many precision DAS applications, however, require differential or two-input amplifiers.

Basic operational amplifiers have both positive and negative input terminals and are therefore capable of such operation. Indeed, they are used in this manner. The simplest of the op-amp differential circuits can be found in Figure 6-14. The circuit is a combination of both the inverting and noninverting op-amp connections. First, place a ground at V_2 and consider the inverting path provided by V_1. The expression for the inverting gain is:

$$A_I = \frac{V_{OI}}{V_1} = -\frac{R_2}{R_1} \qquad (6\text{–}31)$$

Now ground V_1 and consider the noninverting path provided by V_2:

$$A_{NI} = \frac{V_{ONI}}{V_2} = \underbrace{\left[\frac{R_2}{R_1 + R_2}\right]}_{\substack{\text{Caused by } R_1/R_2 \text{ input} \\ \text{attenuation at } V_2 \text{ input}}} \left[\frac{R_1 + R_2}{R_1}\right] = \frac{R_2}{R_1} \qquad (6\text{–}32)$$

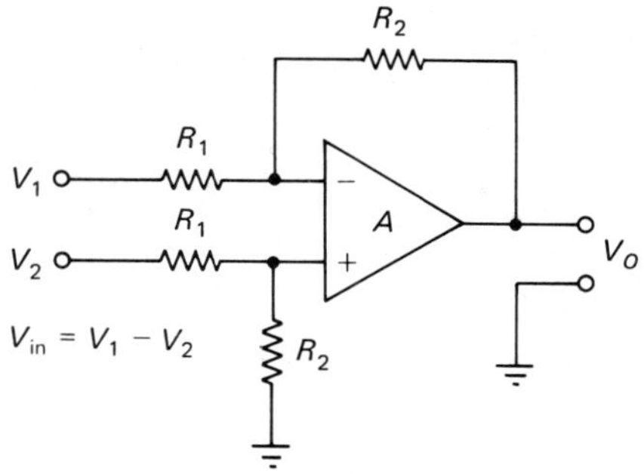

Figure 6–14. Differential Circuit with One Amplifier

If we now consider that the differential output voltage is:

$$V_{OD} = V_{OI} + V_{ONI} = -\frac{R_2}{R_1} V_1 + \frac{R_2}{R_1} V_2 \qquad (6\text{–}33)$$

$$= \frac{R_2}{R_1} (V_2 - V_1) \qquad (6\text{–}34)$$

Unfortunately, this configuration loses many of the good features provided by the two single-input circuits, such as the inverting amplifier's ability to operate with high input voltages and the noninverting circuit's ability to provide high input impedance while providing high gain and wide bandwidths. It does, however, have an advantage over the multiple amplifier configurations of Figures 6–15 and 6–16: Only one set of drift and noise parameters is present. The potential for common-mode problems are also inherent in this configuration. If a common-mode voltage were present at both inputs, the expression for the output voltage error would be:

$$V_{OCM} = -V_{CM}\frac{R_2}{R_1} + V_{CM}\left[\frac{R_2}{R_1 + R_2}\right]\left[\frac{R_1 + R_2}{R_1}\right] \qquad (6\text{–}35)$$

It is imperative that all R_1 resistors be equal to each other and that all R_2 resistors be equal to each other, or V_{OCM} will not be zero.

Figure 6–14 does not contain the source resistances, which must also be balanced. Matching all the resistances is very difficult in actual practice. The common-mode problem can also make gain changing impossible. As we have seen, a change in any resistor value requires all

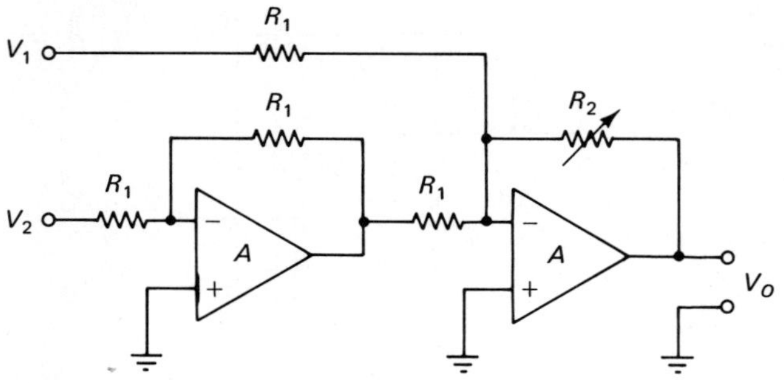

Figure 6-15. Differential Circuit with Two Inverting Amplifiers

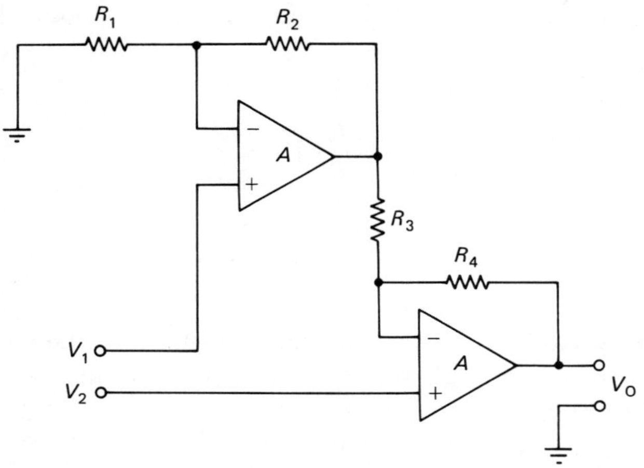

Figure 6-16. Differential Circuit with Two Noninverting Amplifiers

other resistors (R_1 or R_2) to be changed a like amount. Even with these problems, the circuit may be useful for single-gain applications, or in low-noise environments.

Two alternate differential amplifier configurations can be found in Figures 6-15 and 6-16. Each utilizes two separate op-amps. The gain expression for the circuit of Figure 6-15 is:

$$A = \frac{V_O}{V_{in}} = \frac{R_2}{R_1} \qquad (6\text{–}36)$$

where $V_{in} = V_2 - V_1$

The advantages of this circuit are that the op-amps may be inverting-only types; they may be chopper-stabilized, and the gain may be easily changed by varying R_2. Several disadvantages, however, are inherent. First, good common-mode rejection requires that all the R_1 resistors be matched. Second, the offset is four times that of the single-amplifier configuration, as seen in the expression:

$$V_{O\text{-offset}} = \left[1 + 2\frac{R_C}{R_1}\right] V_{OS2} + 2\frac{R_2}{R_1} V_{OS1} \quad (6\text{–}37)$$

Finally, the input resistance is restricted to low values, to preserve the circuit bandwidth.

The gain expression for the circuit of Figure 6–16 is:

$$A = \frac{V_O}{V_2 - V_1} = \left[1 + \frac{R_4}{R_3}\right] \quad (6\text{–}38)$$

providing the ratio $\frac{R_1}{R_2} = \frac{R_4}{R_3}$ be satisfied. This requirement must be satisfied to maintain good common-mode rejection.

Unlike the circuit of Figure 6–15, this configuration provides a high input impedance. It is, however, not desirable if continuously variable gain is desired.

6-6 THE DATA AMPLIFIER

The data amplifier circuit configuration in Figure 6–17 provides high input impedance, high common-mode rejection, and wide gain ranging. The gain for the data amplifier is expressed as:

$$A = \frac{V_O}{V_2 - V_1} = -\frac{R_5}{R_4}\left[1 + \frac{R_1 + R_3}{R_2}\right] \quad (6\text{–}39)$$

Obviously, R_2 is the ideal selection for changing gain; R_5 is not a good choice because it can cause potential common-mode problems for the gain B stage. The overall gain may also be expressed as gain A × gain B, and the common-mode rejection ratio as:

$$\text{CMRR} = \frac{\text{Gain A}\,[1 + \text{Gain B}]}{K} \quad (6\text{–}40)$$

where K = the tolerance of R_4 and R_5.

Intuitively, it seems that the best CMRR would be achieved if gain A

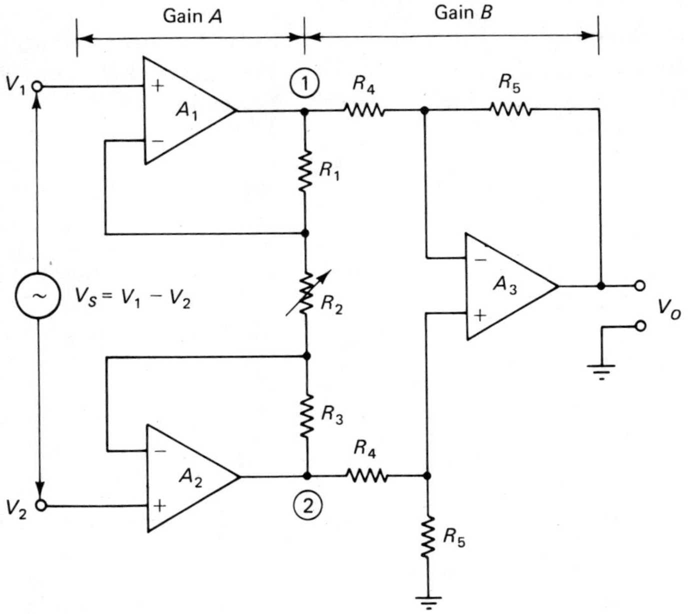

Figure 6-17. The Data Amplifier

were made large and gain B small. For example, if $A_{\text{total}} = 1{,}000$, we would make gain A = 10,000 and gain B = 0.1, rather than A = 0.1 and B = 10,000. Although the direction is correct, there is a limitation to this line of thought. Return to Figure 6-17 and develop the expressions for the voltages at points 1 and 2 ($V_{(1)}$ and $V_{(2)}$), with first V_2 grounded and then V_1 grounded:

With V_2 grounded:

$$V_{(1)} = \left[\frac{R_1 + R_2}{R_2}\right] V_1 \qquad V_{(2)} = -\frac{R_3}{R_2} V_1$$

With V_1 grounded:

$$V_{(2)} = \left[\frac{R_2 + R_3}{R_2}\right] V_2 \qquad V_{(1)} = -\frac{R_1}{R_2} V_2$$

The total expressions for $V_{(1)}$ and $V_{(2)}$ are:

$$V_{(1)} = V_1 + \frac{R_1}{R_2}(V_1 - V_2) \qquad (6\text{–}41)$$

and

$$V_{(2)} = V_2 + \frac{R_3}{R_2}(V_2 - V_1) \qquad (6\text{–}42)$$

Under common-mode conditions, V_1 and V_2 both equal V_{CM}. Factoring this into expressions (6–41) and (6–42), we find that $V_{(1)} = V_{(2)} = V_{CM}$.

Two points are clear here. First, the differential error, due to a common-mode input voltage, is independent of the values or tolerances of the resistors, R_1, R_2, and R_3. Secondly, the voltage range of the amplifiers is affected by the amplitude of the input common-mode voltage as V_{CM} appears at the outputs of both A_1 and A_2. This restricts users from being cavalier with their gain assignments of gain A and gain B. Although the data amplifier is ideal for low-level signals in high noise environments, its bandwidth is generally limited to frequencies in the 200 to 300 kHz range.

6–7 THE SAMPLE-AND-HOLD

Next to basic amplification, the most often used signal conditioning function is that of the sample-and-hold. The sample-and-hold (S/H) circuit is used to take a narrow sample (or "snapshot") of a fast changing input signal and to hold that sample for proper analog-to-digital conversion. The typical S/H has unity gain and is noninverting. In actuality, most S/H circuits are track-and-hold circuits with the sampling taking place when the circuit goes from the track to hold mode. This principle is clearly shown in Figure 6–18: The signal is tracked during the T period and then held during the H period.

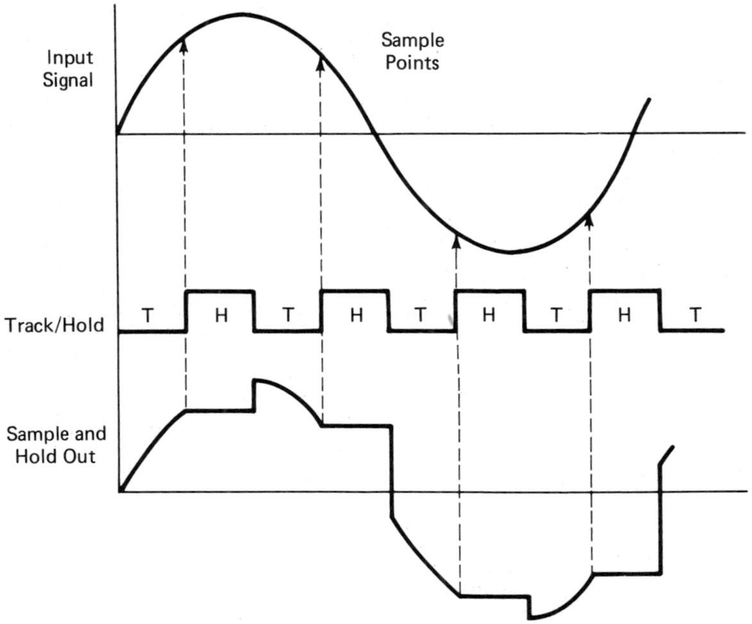

Figure 6–18. Perfect Sample-and-Hold Operation

The basic circuit for realizing the S/H function can be found in Figure 6-19. The switch S_1 must be a fast, high-quality analog device. It must have very precise turn-off time, since that is when the sampling action takes place. S/H circuits with subnanosecond apertures often utilize hot-carrier, quad-diode switches. The capacitor, C, must be a precision capacitor, with very low leakage characteristics, while the amplifier, A, must present a very high input impedance and a very low input current. Both factors help produce a long, stable hold time.

The S/H circuit has four modes of operation:

1. sample or track,
2. track to hold,
3. hold, and
4. hold to track.

Each of these is examined, and the problems and parameters peculiar to each defined.

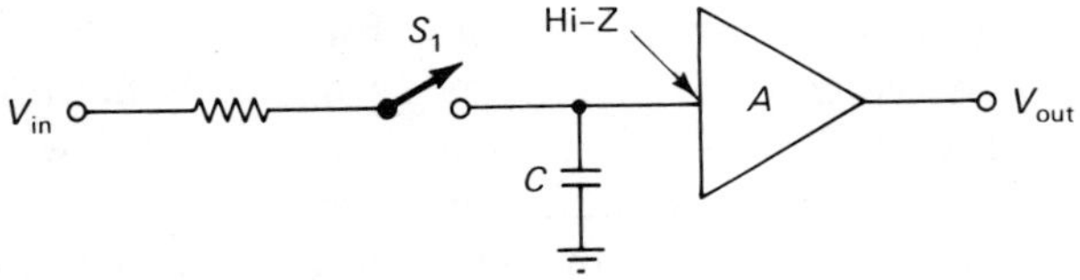

Figure 6-19. Basic Sample-and-Hold Circuit

6-7.1 The Sample (or Track) Mode

Problems encountered while the S/H circuit is in the track mode can be seen in Figure 6-20. In this mode, the S/H circuit is simply an amplifier—subject to all the associated problems, such as phase shift, slew rate, settling time, gain error, offsets, nonlinearity, small signal bandwidth, full power bandwidth, delay, and common-mode error. Analysis for this mode is identical to that used for any amplifier.

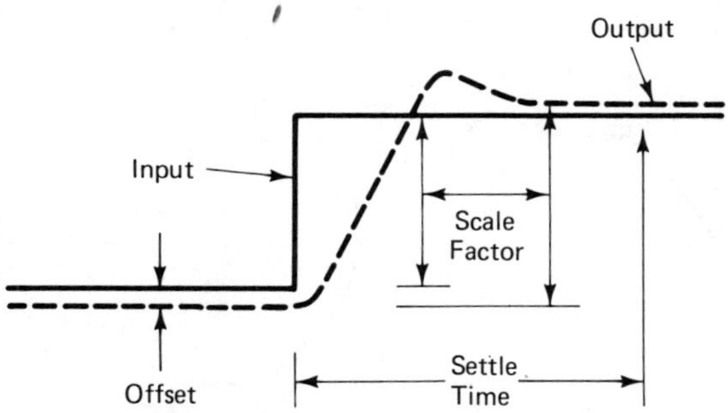

Figure 6-20. Sample, or Track, Mode Problems

6-7.2 Track-to-Hold Mode

Track-to-hold mode problems can be seen in Figure 6-21. The most well-known factor encountered in this mode is aperture time. As we learned in Chapter 2, aperture time is the actual window that occurs at the point of switch opening. It is defined by either of the following expressions:

$$\tau_\alpha = \frac{1}{2^n \pi f} \tag{6-43}$$

or:

$$\left.\frac{dv}{dt}\right|_{max} = \frac{V_{FS}}{2^n \tau_\alpha} \tag{6-44}$$

Actually, aperture consists of two parts: (1) the window that occurs when the switch is opened and (2) the timing jitter due to the uncertainty from sample to sample. If the aperture window is small relative to the period of the highest frequency in the signal being sampled and if the jitter is small, the jitter or aperture uncertainty can be used in place of aperture time in preceding expressions.

Another track-to-hold mode problem is a voltage offset caused by the control circuit dumping charge into the hold capacitor. In certain cases, this problem can be dealt with by adding an out-of-phase signal at the switching time. Switching transients and their settling times must also be reckoned with. Conversion cannot begin until the transient effect has settled to within the desired accuracy.

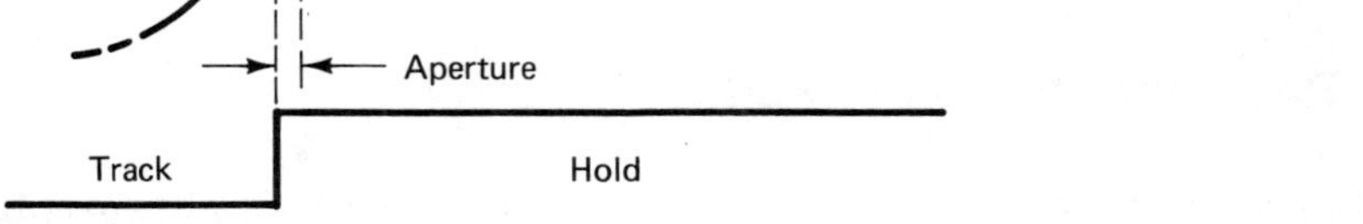

Figure 6-21. Track-to-Hold (Sample) Problems

6-7.3 Hold Mode

The hold mode situation can be seen in Figure 6-22. During this phase of S/H operation, the signal must be maintained within system accuracy long enough for the conversion to take place. The two basic

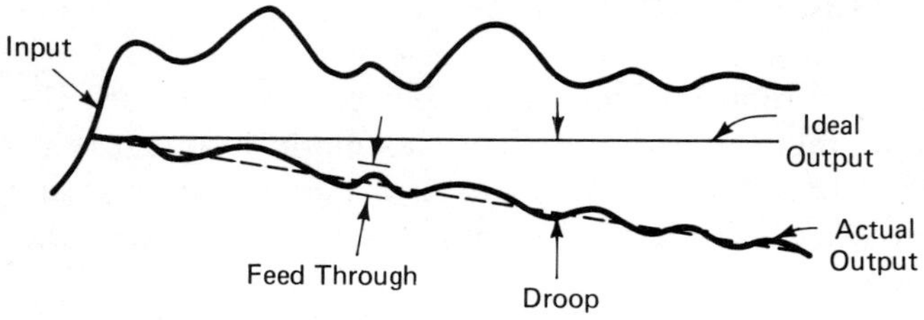

Figure 6-22. Hold Mode Problems

problems that characterize this mode are voltage droop and feed-through.

Droop is the drift or decay of the charge on the hold capacitor. The loss of charge is caused by off-switch leakage, by buffer amplifier offset and bias currents, and by leakage of the hold capacitor itself. For the period of concern, droop generally occurs at a constant rate and can be defined by:

$$\text{Droop} = \frac{dv}{dt} = \frac{I_L}{C_{\text{hold}}} \qquad (6\text{-}45)$$

The *feed-through* problem is caused by a fraction of the input signal feeding through the closed switch interelement capacitance. This factor, usually measured by applying a full-scale sinusoidal input, is expressed in terms of dB over a band of frequencies.

One other factor that should be considered for the very fast sample-and-hold circuit is dielectric absorption. Dielectric absorption describes a phenomenon that occurs after the hold capacitor has accepted a large charge, very fast. The held value tends to "creep" after the circuit is in the hold mode for a while. This "creep" is caused by charge redistribution within the capacitor, and it depends on the capacitor type. This effect is less than 0.01 percent for good teflon or polystyrene capacitors.

6-7.4 Hold-to-Track Mode

Hold-to-track mode problems can be seen in Figure 6-23. In this phase of S/H operation, the hold capacitor must acquire new voltage that the input has reached while the circuit was in the hold mode. Acquisition time is the name given to this parameter. The worst case condition exists when the hold capacitor is at $-V_{FS}$ and the input has reached $+V_{FS}$, or vice versa. The minimum track time is determined by the time needed to reach and settle out to the new value.

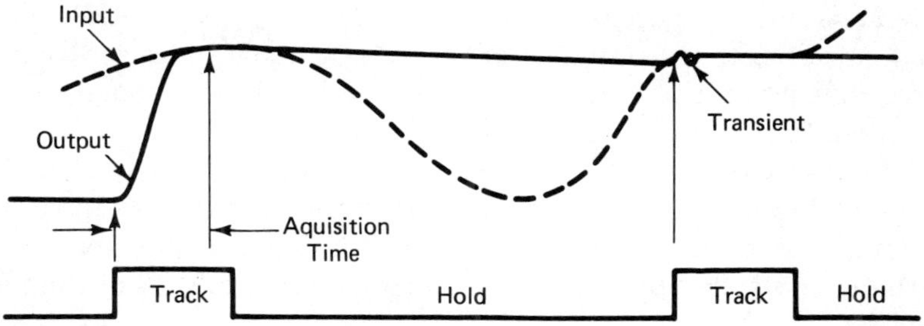

Figure 6–23. Hold-to-Track Mode Problems

As switching is involved, this mode is also plagued with switching transients. This condition is most troublesome when the newly acquired value is very close to the previously held value as seen on the right-hand side of Figure 6–23. The value that remains on the hold capacitor can be as much a function of the transient as it is of the new input value.

6–7.5 Sample-and-Hold Implementations

The simplest implementation of a sample-and-hold circuit can be found in Figure 6–24. This circuit, which does not have a buffer amplifier before the switch, is typically used when the signal source is buffered, or when the analog multiplexer contains an amplifier. When the switch is closed, C_H is charged through the source resistance R.

The acquisition time is a function of the source resistance and the source's current drive capability. When the capacitor has charged to the required accuracy, the switch may be opened. Note should be made that in cases where R_S is small, settling time at the capacitor may be less than that observed at the amplifier output. In this situation, the amplifier may have the limiting bandwidth. It is not uncommon to use

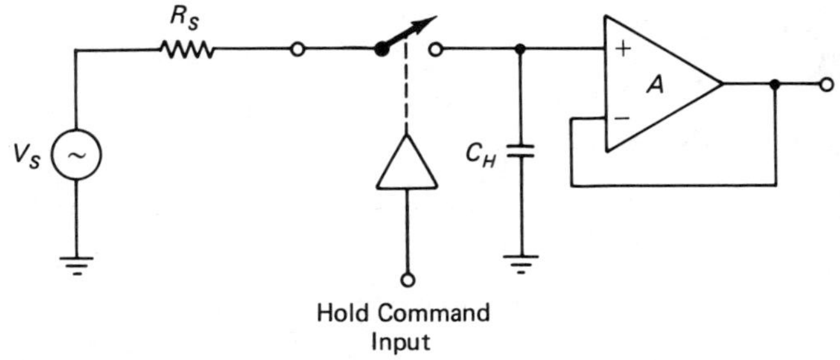

Figure 6–24. The Simplest Sample-and-Hold Circuit

the S/H circuit of Figure 6-24, if the user does not wish to add an additional amplifier to the system. Care should be taken, however, that the source that drives the S/H circuit can handle a large capacitor on its output without oscillating.

The same type S/H circuit can be found with a built-in prebuffer, as seen in Figure 6-25. Operation of this circuit is similar to the one previously discussed, with the exception that the acquisition time is internally controlled by C_H and the output characteristics of the input amplifier.

Where high accuracy is required, more elaborate circuits, such as that seen in Figure 6-26 are utilized. In this case, the hold capacitor C_H is contained within a feedback loop during the acquisition cycle (S_1 closed, S_2 open). The accuracy of the voltage on C_H and its linearity over the operating range is thereby improved by the loop gain. During the hold mode (S_1 open, S_2 closed), the loop is broken. Amplifier A_2 becomes a simple high impedance output buffer, while A_1 simply tracks the S/H input. The closure of S_2 provides A_1 with its own feed-

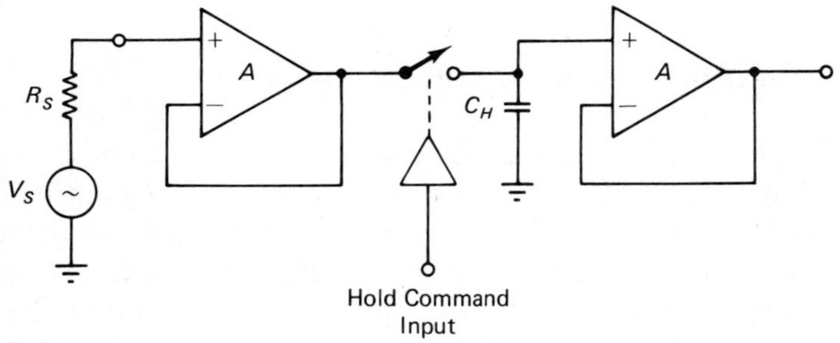

Figure 6-25. Sample-and-Hold with Input Buffering

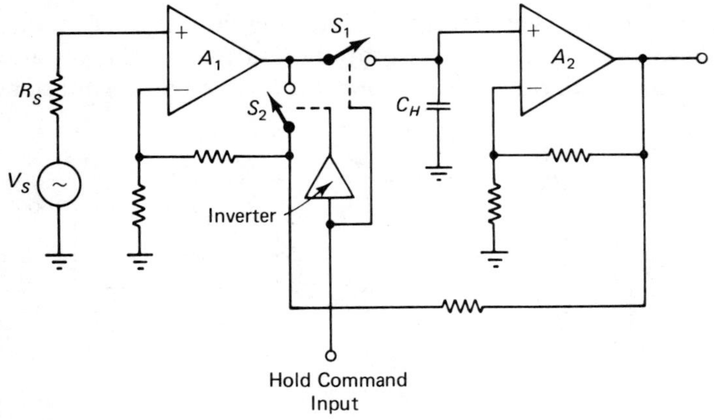

Figure 6-26. Sample-and-Hold with Feedback

back path during the hold mode. Sample-and-hold units that employ feedback achieve fairly high sampling accuracy, at the expense of acquisition time.

6–8 THE PEAK-DETECT-AND-HOLD

A derivative of the sample-and-hold is the peak-detect-and-hold circuit. The main goal of this circuit is the capture of the highest peak signal, which occurs during a given time interval, and to hold it so that an accurate analog-to-digital conversion can be made. Figure 6–27a contains a basic peak-detect-and-hold circuit. Whenever the input voltge, V_S, exceeds that on the hold capacitor, the output of A_1 is connected to the capacitor via the diode, D_1. The voltage on the capicitor then tracks V_S (Figure 6–27b). When V_S starts to decrease, the diode ceases to conduct, and the peak value is held on C_H. The clamp is usually a fast diode, which prevents the amplifier output from going into saturation. After the conversion is completed, S_1 is closed, momentarily discharging C_H. When the capacitor is discharged, the circuit is ready to respond to the next peak.

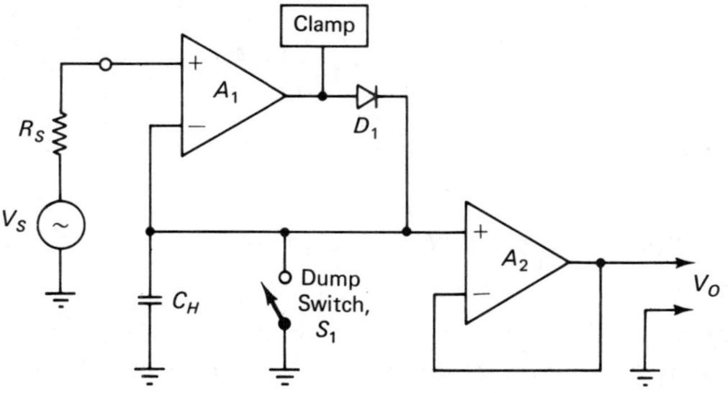

(a) The basic circuit

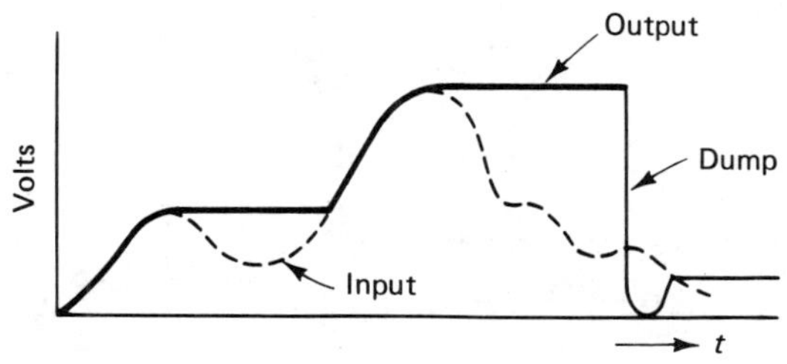

(b) Waveforms

Figure 6–27. Peak-Detect-and-Hold Circuit

6-9 OTHER SIGNAL CONDITIONERS

Two other signal conditioners that have not yet been discussed are nonlinear amplifiers and active filters. The most usual nonlinear amplifier configuration is an operational amplifier with a nonlinear element in the feedback loop, as shown in Figure 6-28. The nonlinear device most often used is a bipolar silicon transistor operated at low levels.

Typical transfer functions might be:

$$\text{Log current:} \quad V_{out} = -A \log_{10} I_{in} \qquad (6\text{–}46)$$

$$\text{Log voltage:} \quad V_{out} = -A \log_{10} V_{in} \qquad (6\text{–}47)$$

$$\text{Antilog:} \quad V_{out} = A \times 10^{(-1/B)V_{in}}$$

Two simple active filters are found in Figure 6-29. The simple integrator of Figure 6-29a is widely used in slow-speed analog-to-digital converters and, as you would expect, has the following time response:

$$V_O = -\frac{1}{RC} \int V_S dt \qquad (6\text{–}48)$$

If offset current and voltage are considered, the expression becomes:

$$V_O = -\frac{1}{RC} \int V_S dt + \frac{1}{RC} \int V_{OS} dt + \frac{1}{C} \int I_{OS} dt \qquad (6\text{–}49)$$

Figure 6.29b shows a simple band-pass filter configuration. Many, many more circuits are available, simulating almost every type of analog filter. Because the subject is so broad, it is only introduced here. Should you desire to use an active filter, consult the available texts that are heavily devoted to this area (References 5 and 6).

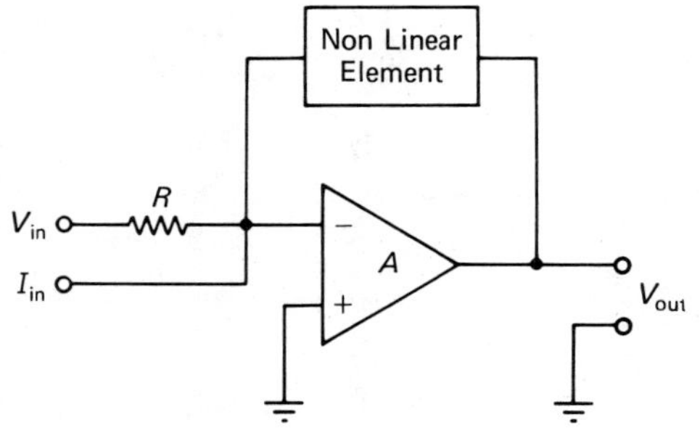

Figure 6-28. Nonlinear Amplifier

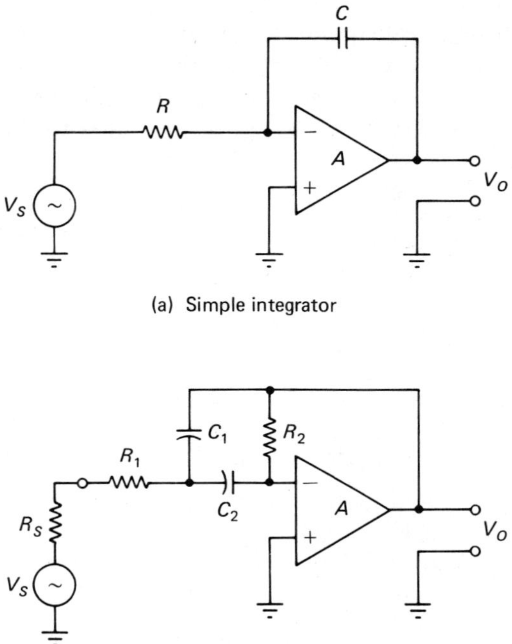

Figure 6-29. Active Filters

PROBLEMS

6–1. An inverting operational amplifier has the following characteristics:

$$R_S = 1\ \text{k}\Omega \quad A_{OL} = 500 \quad I_{OS} = 0.1\ \text{nA}$$
$$R_I = 9\ \text{k}\Omega \quad I_b = 1\ \text{nA}$$
$$R_F = 50\ \text{k}\Omega \quad V_{OS} = 10\ \mu\text{V}$$

Calculate:

a. A in a closed loop,
b. R_{in} in a closed loop,
c. ΔV_O due to V_{OS}, and
d. ΔV_O due to I_b.
e. If a compensating resistor, R_C, were added, what would its value be and what would ΔV_O due to I_{OS} be?

6–2. An amplifier has a full-power bandwidth of 5 mHz, a full voltage swing of 10 V pp, a slew rate of 500 V/μsec, and a start-up delay of 40 nsec. What is the settling time to 0.1 percent?

6–3. An inverting amplifier has an input resistance ($R_I + R_S$) of 100 kΩ and a feedback resistor R_F of 500 kΩ. The stray capacitance across the feedback resistor is 5 pF. The Gain × Bw product of the amplifier is 10 MHz. What is the closed-loop bandwidth of this amplifier?

6–4. An amplifier, connected in a noninverting configuration, has the following characteristics:

$R_S = 1$ kΩ	$V_{OS} = 10\ \mu V$	$R_{in\text{-}OL} = 100$ kΩ
$R_1 = 25$ kΩ	$I_b = 1$ nA	
$R_2 = 5$ kΩ	$A_{OL} = 180$	

Calculate:
a. A in a closed loop,
b. R_{in} in a closed loop,
c. ΔV_O due to V_{OS}, and
d. ΔV_O due to I_b. (Note: Ignore R_{in}-common mode.)
e. How could you improve the effect the ΔV_O due to I_b?

6–5. An operational amplifier has the following noise characteristics:

$$e_n = 2\ \mu V\ (RMS)$$
$$i_n = 5\ pA\ (RMS)$$

It is connected in an inverting configuration with:

$R_1 = 49$ kΩ	$R_C = 45$ kΩ
$R_S = 1$ kΩ	$\Delta f = 10$ kHz
$R_F = 450$ kΩ	

What is the noise effect at the amplifier output?

6–6. Using the amplifier of problem 6–5 in a noninverting connection with:

$R_1 = 80$ kΩ	$R_S = 1$ kΩ
$R_2 = 10$ kΩ	$\Delta f = 10$ kHz

what is the noise effect at the amplifier output?

6–7. A differential amplifier with a gain of 100 and a common-mode rejection of 100 dB at 60 Hz is subjected to an environment with a common-mode voltage of 5 V at 60 Hz. What is the output of the amplifier due to V-common-mode?

6–8. A data acquisition system utilizes a single-amplifier differential circuit as shown in Figure 6–14. In this case, $R_1 = 1{,}000\ \Omega$ and $R_2 = 2{,}000\ \Omega$. Consider that the R_1 in the V_2 leg is high by 0.1 percent and $V_{CM} = 10$ V. What is V_{OCM} under these conditions?

6–9. The data amplifier of Figure 6–17 is used in a data acquisition system. If R_1 and $R_3 = 5$ kΩ, $R_2 = 101\ \Omega$, $R_4 = 10$ kΩ, $R_5 = 20$ kΩ and if the tolerance of R_4 and R_5 is 0.01 percent, what are the overall gain and the amplifier CMR?

6–10. A sample-and-hold circuit is utilized in a data acquisition system. The input full-scale voltage is $+10$ V. The highest frequency to be converted is 1 kHz, and the sample rate is 10 kHz (1 sample/cycle). The A/D converter conversion time is 10 μsec. The selected S/H has an aperture time of 20 nsec, a droop rate of 100 μV/μsec, and an acquisition time of 50 μsec.

a. If the allowable aperture error and droop error are each 0.02 percent, can the S/H meet these requirements?
b. Is the acquisition time satisfactory?

REFERENCES

1. G. E. Tobey, et al, *Operational Amplifiers—Design and Application*. New York: McGraw-Hill Book Company, 1971.
2. J. I. Smith, *Operational Circuit Design*. New York: Wiley-Interscience, 1971.
3. D. H. Sheingold, *Analog–Digital Conversion Handbook*. Norwood, Mass.: Analog Devices, Inc., 1976.
4. R. Demrow, *Evolution from Operational Amplifier to Data Amplifier*. Norwood, Mass.: Analog Devices Technical Bulletin, 1968.
5. P. H. Garrett, *Analog Systems for Microprocessors and Minicomputers*. Reston, Va.: Reston Publishing Company, Inc., 1978.

7 Digital-to-Analog Converters

7-0 INTRODUCTION

The digital-to-analog (D/A) converter is used to reconstitute the signal after it has been converted to digital form and processed. The basic D/A converter consists of a group of weighted resistors that are interconnected to produce an analog voltage to correspond with a given digital input.

Why are we discussing a data acquisition output device in the midst of a series of chapters on DAS input? In the next chapter, we will discuss the analog-to-digital (A/D) converter. As we will see, many A/D converters contain a D/A converter as a part of the conversion mechanism. In other words, you need a pre-knowledge of the digital-to-analog conversion operation.

A block diagram of the typical D/A converter can be seen in Figure 7-1. It contains the weighted resistor (ladder) network, a set of precision analog switches, and a storage register to hold the digital quantity being converted. If the converter is a high-quality, high-accuracy unit, it also contains a precision voltage reference. The amplifier provides buffering between the D/A converter and the element using the output.

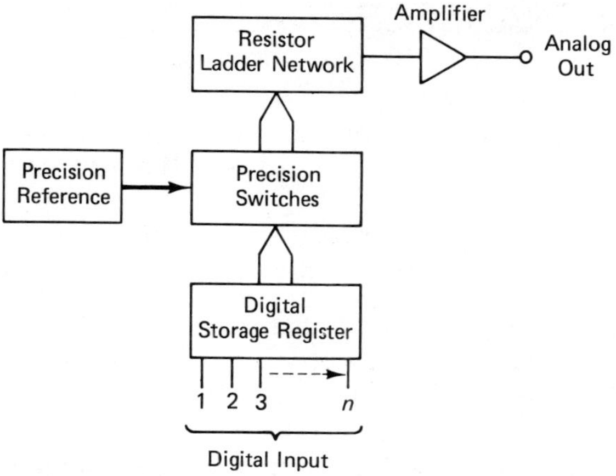

Figure 7-1. Digital-to-Analog Converter Block Diagram

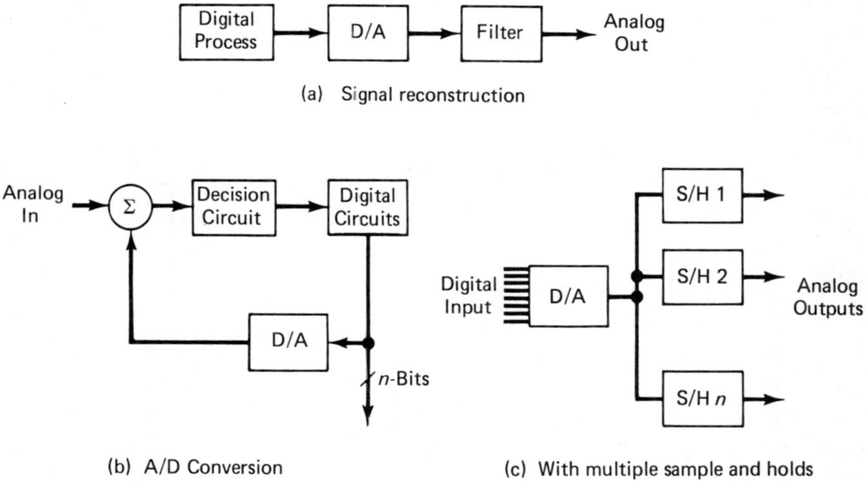

Figure 7-2. Typical Uses for the D/A Converter

The most recognized use of the D/A converter is signal reconstruction, as seen in Figure 7-2a. Here the D/A converter is followed by a filter whose function is to smooth out the steps. Figure 7-2b shows a D/A converter as part of an A/D converter—a not uncommon situation.

Another use of the D/A converter is seen in Figure 7-2c where a single D/A feeds multiple outputs via a series of sample-and-hold (S/H) units. The D/A is multiplexed among the several channels.

7-1 THE TRANSFER FUNCTION

The transfer function for a 4-bit digital-to-analog converter can be found in Figure 7-3. This function is similar to the quantizer transfer function found in Figure 3-1. Two major differences can be seen:

1. The input and output axes have been reversed, and now the horizontal is Binary Code In and the vertical is Analog Voltage Out.
2. Close observation also shows that the first step is a full quantum step, rather than the half step of Figure 3-1. For the D/A converter, the output voltage must exactly represent the digital code, while the quantizer of Figure 3-1 must make a decision, so that the output code centers itself about the analog voltage it represents.

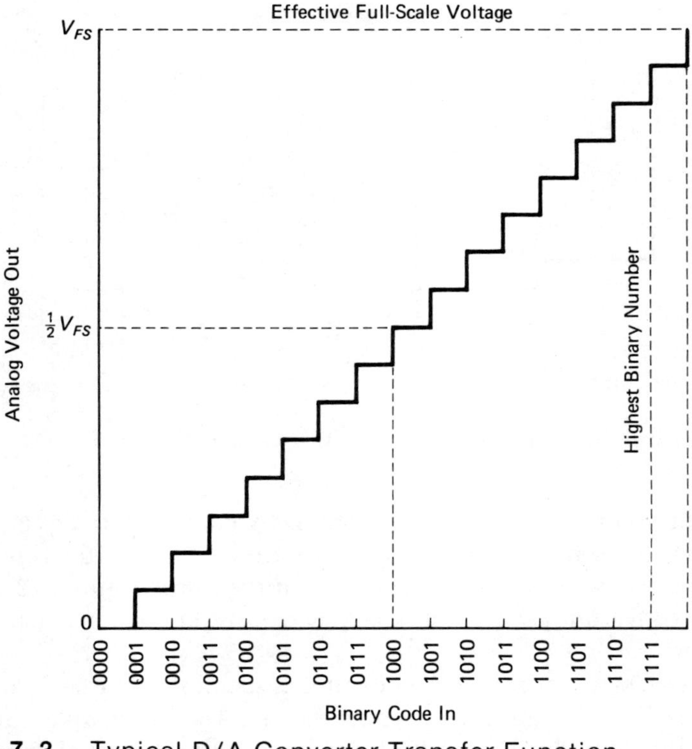

Figure 7-3. Typical D/A Converter Transfer Function

The major system parameters, which are of concern to the D/A converter, are as follows:

1. absolute accuracy,
2. relative accuracy,
3. linearity,
4. monotonicity,
5. temperature coefficients,
6. settling time, and
7. conversion rate.

The first six of these parameters are fully discussed in Chapter 4. The last, *conversion rate*, is the rate at which the D/A converter can accept new digital inputs, while still allowing time to use the outputs after settling. Typically, conversion rates range from a few hundred nanoseconds to a few tens of microseconds, depending on the number of bits being converted and the quality of the converter.

We will now discuss some of the most popular D/A converter implementations.

7-2 THE WEIGHTED RESISTOR D/A CONVERTER

One of the two most basic D/A converter circuits is the weighted resistor or current summing network of Figure 7-4. The network consists of a series of binary weighted resistors, all tied to a common summing point. Close inspection shows that each successive resistor doubles in value when moving from the most significant bit (MSB) to the least significant bit (LSB). When a given bit input is a logical one, the associated resistor is connected to the reference (E_{ref}) by a precision switch.

The network tends to look like a current source with an equivalent resistance which may be represented by:

$$R_T = \frac{2^{(n-1)}}{2^n - 1} R \qquad (7\text{-}1)$$

where R = the value of the MSB resistor

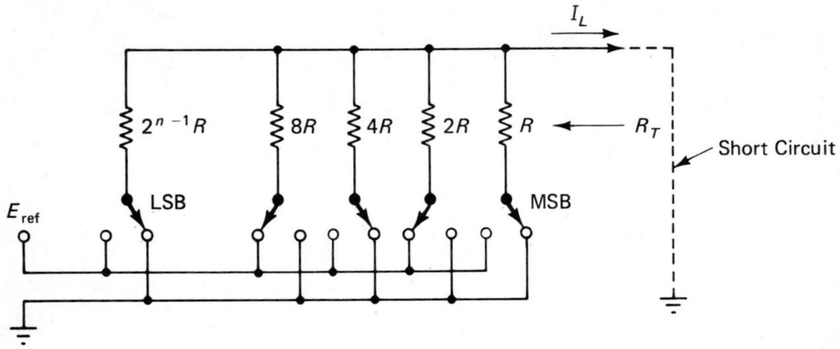

Figure 7-4. Weighted Resistor or Current Summing Network

For a 4-bit converter:

$$R_T = \frac{2^3}{2^4 - 1} R = \frac{8}{15^R} = 0.533R$$

and for a 10-bit converter:

$$R_T = \frac{2^4}{2^{10} - 1} R = \frac{512}{1{,}023} R = 0.500R$$

You can see that, as the number of bits increases, the equivalent resistance approaches $R/2$.

The output current, I_L, is described by:

$$I_L = E_{\text{ref}} \sum_{n=1}^{n} \frac{a_k}{2^{(n-1)}R} \qquad (7\text{–}2)$$

where the a's = either one or zero

The maximum current is:

$$I_{L_{\max}} = \frac{E_{\text{ref}}}{R_T} \qquad (7\text{–}3)$$

Note that, when a resistor is not used, it is connected to ground. This maintains the equivalent network resistance at R_T at all times. If unused resistors are not tied off, and if the circuit network it is driving is not of a low impedance, substantial nonlinearity could result.

The weighted resistor network works well in combination with an inverting operational amplifier. This circuit, shown in Figure 7-5, has the network connected directly to the virtual ground summing point.

The results, translated to the output, are:

$$V_O = -I_L R_F \tag{7-4}$$

or, substituting expression (7-2),

$$V_O = -\left[E_{\text{ref}} \sum_{n=1}^{n} \frac{a_k}{2^{(n-1)}R}\right] R_F \tag{7-5}$$

where the a's = either one or zero

At full scale, substituting expression (7-3):

$$V_O = -\frac{E_{\text{ref}}}{R_T} R_F \tag{7-6}$$

or, substituting expression (7-1):

$$V_O = -E_{\text{ref}} \left[\frac{2^n - 1}{2^{(n-1)}R}\right] R_F \tag{7-7}$$

Consider a 10-bit system, where $R_F = R/2$:

$$V_O = -E_{\text{ref}} \left[\frac{1{,}023}{512R}\right] \frac{R}{2}$$

$$= -\frac{1{,}023}{1{,}024} E_{\text{ref}}$$

We should not be surprised to find that the result is one quantum level below the effective full-scale range shown in Figure 7-3.

An alternate configuration for the weighted resistor network can be found in Figure 7-6. In this circuit, the precision switches are on the opposite end of the ladder network. The main advantage of this connection is that the switched ends of the resistors always see the same condition—either real ground or a virtual ground. Such being the case, the respective currents through the various resistors are also constant. This substantially reduces the "glitches" or switching transients.

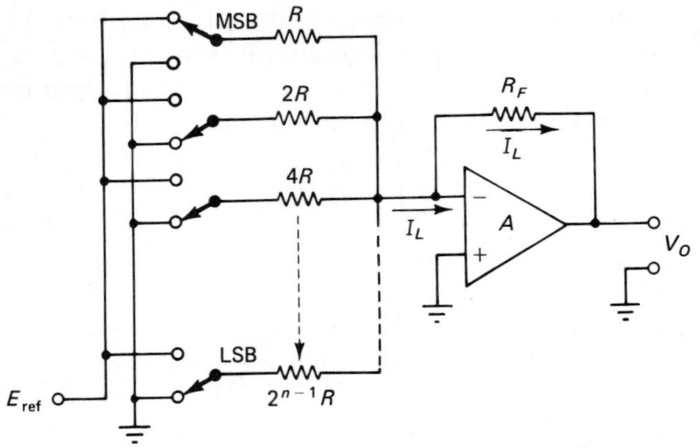

Figure 7-5. Weighted Resistor Network with Op-Amp

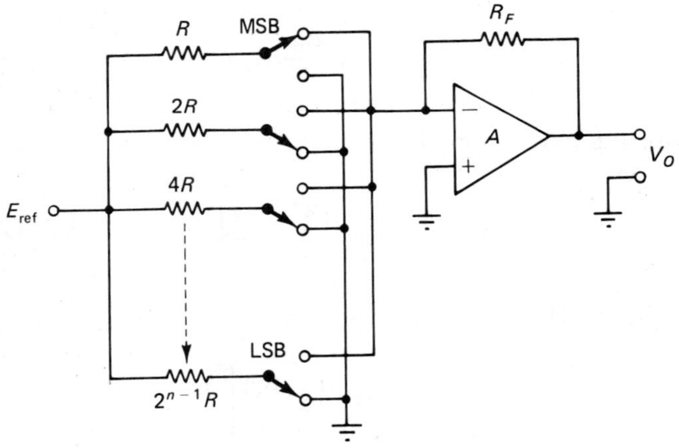

Figure 7-6. Alternate Weighted Resistor Connection

If a sample-and-hold function is added to the circuit, and if sampling occurs after the remaining transient has settled, glitching may be totally eliminated.

An actual implementation of the concept proposed by Figure 7-6 is seen in Figure 7-7. When the digital inputs are all logical zeros, the resistor currents all pass through the diodes, and none flows into the amplifier's input. Whenever an input goes to a logical one, the clamp diode is released, and the resistor current is routed through its associated transistor to the amplifier.

One problem with the weighted resistor network is that it requires a wide range of resistor values for a large number of bits. For example, in a 12-bit ladder, 12 different resistor values are needed. If the MSB resistor is $R\ \Omega$, then the LSB resistor must be $2^{(n-1)}R$, or $2{,}048R\ \Omega$.

Yet large resistance values tend to increase noise and reduce converter operating speed. The customary way to get around this problem is to use multiple identical, subladder sections in a *step-down* configuration. Two examples are found in Figure 7-8. One utilizes a terminating resistor, while the other does not. Consider first the network of Figure 7-8a, which does not have a terminating resistor. The contribution to the total maximum output current (I_L) from sections 1 and 2, respectively, is 16:1. Let's compute the value of the connecting resistor, R_S:

$$I_{L_1} = \frac{15}{16} I_{L_2} \qquad \text{and} \qquad I_{L_2} = \frac{E_{ref}}{8R}$$

therefore:

$$I_{L_1} = \frac{15}{16} \times \frac{E_{ref}}{8R} = \frac{15}{128R} E_{ref}$$

Also, it can be proven that:

$$I_{L_1} = \frac{E_{ref}}{R_P + R_S}$$

where $R_P = R_T$ of a section, or in this case
$R_P = \frac{8}{15} R$

We can now set up the expression:

$$I_{L_1} = I_{L_1}$$

or:

$$\frac{15E_{ref}}{128R} = \frac{E_{ref}}{\frac{8}{15}R + R_S}$$

Solving, we find that:

$$R_S = 8R \tag{7-8}$$

Now consider the network of Figure 7-8b. The expression for the terminating resistor R_{term}, is:

$$R_{term} = R_P(r - 1) \tag{7-9}$$

where $R_P = R_T$ of a section
r = the ratio of the sections

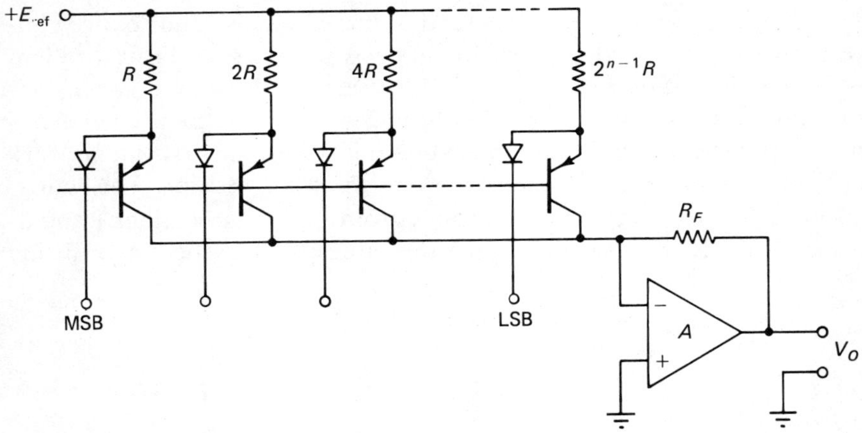

Figure 7-7. Typical Circuit for Weighted Resistor Circuit

For the example shown, R_P is $\frac{8}{15}R$ and r is 16 to 1. The terminating resistor value is therefore:

$$R_{\text{term}} = \tfrac{8}{15}R\,(16 - 1) = 8R$$

Looking to the left into section 2 through R_S, you see that the resistance must be equal to the LSB resistor of section 1, or $8R$. The value of the connecting resistor must then be:

$$R_S + 8R - \frac{R_P R_{\text{term}}}{R_P + R_{\text{term}}}$$

where $R_P = R_T$ of a section

For the example given:

$$R_S = 8R - \frac{\frac{8}{15}R(8R)}{\frac{8}{15}R + 8R}$$

$$= 8R - \frac{R}{2} = 7.5R$$

The solutions provided for Figures 7-8a and 7-8b are unique for two subladder networks. While three and four subladder networks have been used, determination of the connecting and terminating resistors is much more complicated.

Another variation of the weighted ladder is found in Figure 7-9.

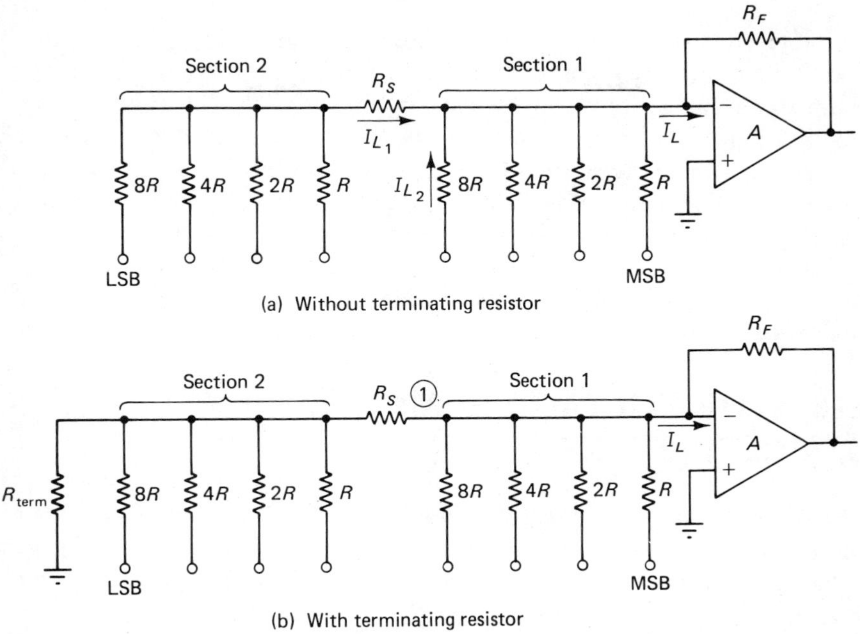

(a) Without terminating resistor

(b) With terminating resistor

Figure 7-8. Weighted Resistor Network with Step-Down

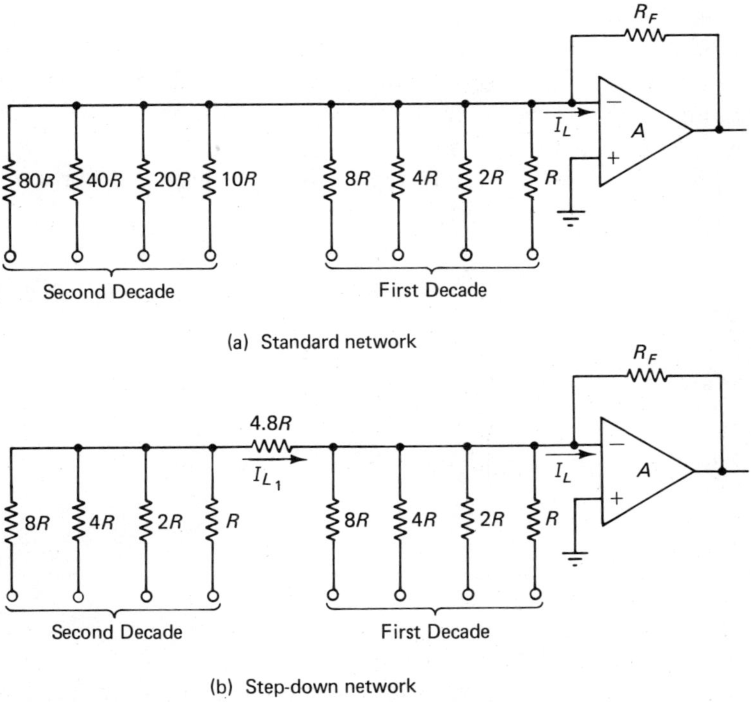

(a) Standard network

(b) Step-down network

Figure 7-9. The BCD Weighted Networks

These circuits cover the situation where the digital input is in binary-coded decimal (BCD) form (Section 3-1). The resistors are grouped in fours, with each group representing a decade. For the standard network of Figure 7-9a, the resistor values for the second decade are ten times larger than those of the first, thus making its contribution ten times less. The same resistor value problem that plagued the binary network is also present in the BCD ladder. Figure 7-9b shows the step-down version of this network. The maximum value of the current I_{L_1} is one-tenth the value of that contributed by the first decade. Verification of the connecting resistor value is left to you.

7-3 THE VOLTAGE SUMMING D/A CONVERTER

The other resistor network commonly used for D/A conversion is the voltage summing or $R/2R$ ladder circuit. This network, seen in Figure 7-10, results if we carry the step-down philosophy of Figure 7-8b to the limit. An important attribute of the $R/2R$ network is that the equivalent resistance R_T is a constant, and:

$$R_T = R \tag{7-10}$$

This statement is easily proven by starting at the LSB side of the network and moving towards the MSB; that is, $2R$ in parallel with $2R$ is R, in series with R is $2R$, and so on.

The maximum output voltage (when the highest binary number is presented to the input) is:

$$E_{O\text{-max}} = \frac{2^n - 1}{2^n} E_{\text{ref}} \tag{7-11}$$

E_0 for any particular binary input is:

$$E_O = E_{\text{ref}} \left[\frac{a_1}{2} + \frac{a_2}{4} + \frac{a_3}{8} + \ldots + \frac{a_n}{2^n} \right] \tag{7-12}$$

or:

$$E_O = E_{\text{ref}} \sum_{n=1}^{n} \frac{a_n}{2^n} \tag{7-13}$$

where a's = either one or zero

If the MSB were a one, and if all the other bits were zero's,

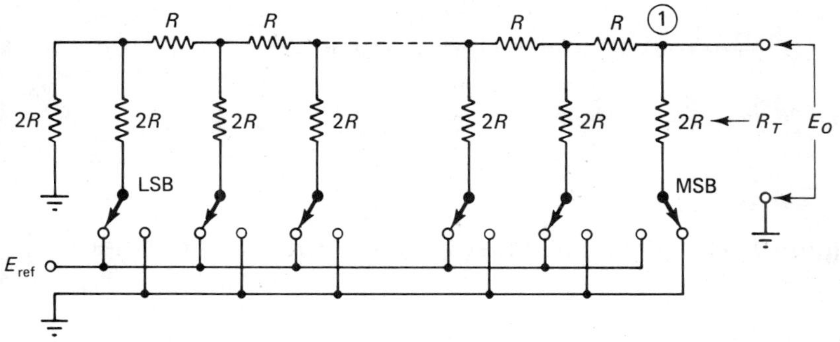

Figure 7-10. The Voltage Summing or *R*/2*R* Ladder Network

expression (7-12) indicates that E_O should equal one-half E_{ref}. This is easily proven. Looking left from point 1 of Figure 7-10, you see that the remainder of the network looks like $2R\ \Omega$ to ground, under these conditions. At the MSB then, E_{ref} sees the equivalent of two series $2R$-Ω resistors to ground. The voltage, E_O, is measured between the center of the two resistors and ground. As the resistors are of equal value, E_O for the MSB is, of course, one-half E_{ref}. Each bit contribution may be analyzed in this fashion.

Other advantages of the $R/2R$ network over the weighted resistor network are:

1. the current ratios in the ladder legs are smaller;
2. only two sizes of resistors are utilized; and
3. the resistors can be lower in value. (The lower value resistors allow a higher-speed operation.)

As the $R/2R$ ladder network is basically a voltage summing circuit, it is only natural that its output be buffered by a high-impedance noninverting op-amp, as seen in Figure 7-11. In this configuration, the network output simply appears at the amplifier output.

Yet, it is also not uncommon to find the $R/2R$ ladder used in conjunction with the inverting configuration shown in Figure 7-12. If the amplifier feedback resistor in this figure is R, the results are the same as with the noninverting amplifier, with the exception that the output is inverted. That is:

$$E_O = -E_{ref} \sum_{n=1}^{n} \frac{a_k}{2^n} \qquad (7\text{-}14)$$

where a's = either one or zero

Consider the situation when the MSB/2 is a logical one and all

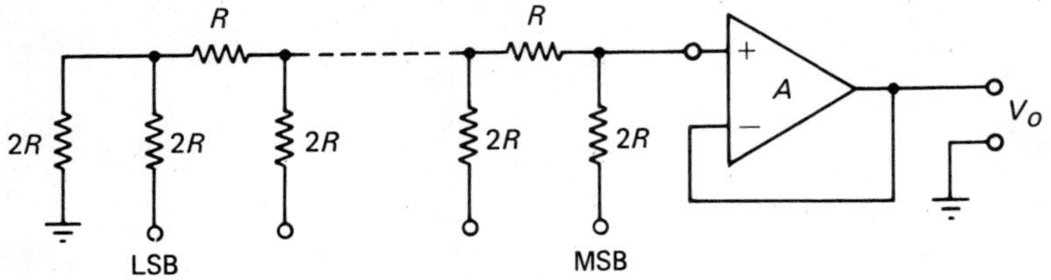

Figure 7-11. Voltage Ladder with Noninverting Amplifier

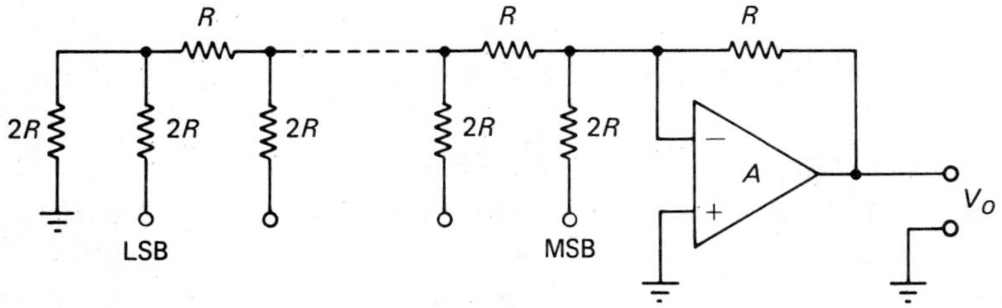

Figure 7-12. Voltage Ladder with Inverting Amplifier

other bits are logical zero's (Figure 7 - 13). With point 1 a virtual ground, the most significant bit leg of the network simply does not appear in the calculation which follows:

$$V_2 = \frac{\frac{2}{3}R}{2R + \frac{2}{3}R} E_{ref} \qquad (a)$$

where $\frac{2}{3}$ is the parallel value of the $\frac{\text{MSB}}{2}$ resistor and R between 1 and 2

$$V_2 = \frac{2R}{2R + 2R} E_{ref} = -\frac{1}{4} E_{ref} \qquad (b)$$

$$E_O = -\frac{R}{R}\left[\frac{1}{4} E_{ref}\right] = -\frac{1}{4} E_{ref} \qquad (c)$$

Contributions from each of the other bits may be analyzed in a similar manner.

This network may also be used in the inverted mode, with the precision switches on the other side of the ladder (Figure 7-14a). As for the current summing case, this circuit connection allows the switches to work between two similar voltages (real ground and virtual ground), and it maintains a constant current through the network. Also, switching transients are reduced and speed improved in this situation.

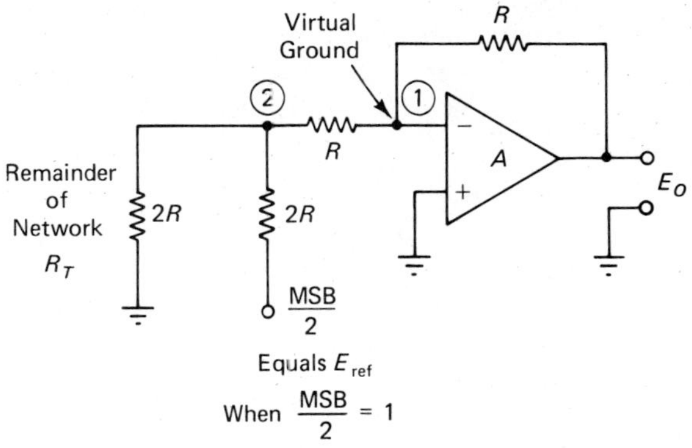

Figure 7-13. *R*/2*R* Network and Inverting Amplifier with Only MSB a Logic 1

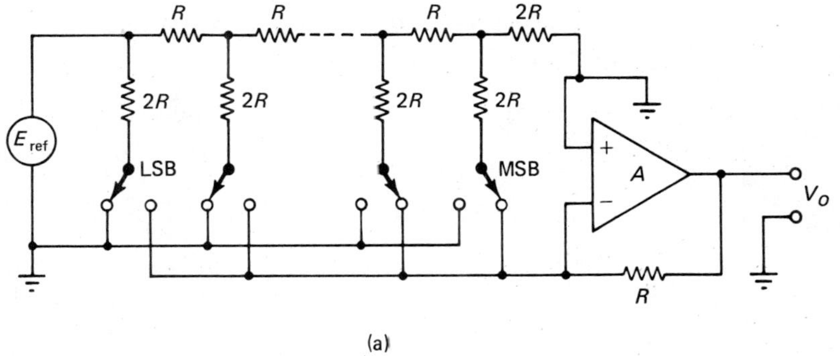

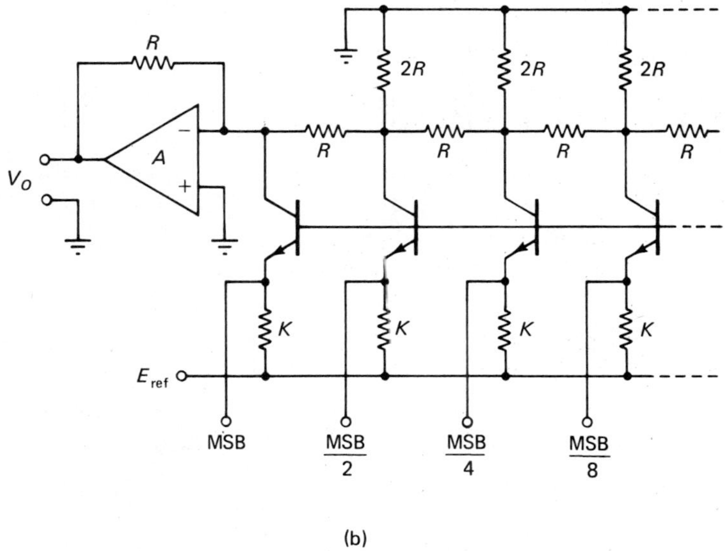

Figure 7-14. Alternate Configurations for the Voltage Ladder

Figure 7-14b provides still another alternative for the $R/2R$ network. When a given bit is a logical zero, its associated transistor is cut off and the current E_{ref}/K is shorted to ground via the bit input connection. When the bit is at a logical one, the transistor is operational and a metered current is pulled from the appropriate leg of the ladder network. This circuit provides high speed and is easy to implement. All transistor base-to-emitter voltages and switch times must be the same. It has the distinct disadvantage that both the ladder network resistors and the K-resistors must be matched.

PROBLEMS

7–1. A 12-bit D/A converter utilizes a current-summing ladder network. Its most significant bit (MSB) resistor value is 1,000 Ω.
 a. What are the values of R_T and the least significant bit R?
 b. What is I_{max} if $E_{ref} = +10$ V?

7–2. The D/A converter of problem 7-1 is connected to the summing point of an inverting operational amplifier.
 a. What is $V_{output\text{-}max}$ if $R_F = 1{,}000\ \Omega$?
 b. What is V_{output} if only the MSB/4 and MSB/8 inputs are at logical ones.

7–3. A ladder network of the form shown in Figure 7-8b is used in a D/A converter. In this case, a 10-bit network is desired and each of two sections contains five resistors.
 a. If the MSB resistor is 1,000 Ω, what are the values of R_S and R_{term}?
 b. What is the total network R_T?

7–4. Prove that the second decade of Figure 7-9 provides one-tenth the contribution of the first.

7–5. A 10-bit D/A converter utilizes a voltage-summing resistor network with R equal to 500 Ω, and E_{ref} equal to +10 V.
 a. What is R_T and what is $E_{output\text{-}max}$?
 b. Show that the MSB/2 produces $E_{ref}/4$ at the output.

REFERENCES

1. D. F. Hoeschele, *Analog-to-Digital/Digital-to-Analog Conversion Techniques*. New York: John Wiley & Sons, Inc., 1968.
2. H. Schmid, *Electronic Analog/Digital Conversions*. New York: Van Nostrand Reinhold Company, 1970.
3. D. H. Sheingold, *Analog–Digital Conversion Handbook*. Norwood, Mass.: Analog Devices, Inc., 1972.

8

Analog-to-Digital Converters

8-0 INTRODUCTION

Analog-to-digital conversion is at the heart of the data acquisition system. It is the interface between the analog and the digital domains. Analog signals are sampled, quantized, and encoded into a digital format. In many cases, all three functions are accomplished within the analog-to-digital (A/D) converter package. If an external sample-and-hold circuit is used, the A/D converter provides only the quantization and encoding functions (Chapter 3). Here, the sample-and-hold circuit maintains the sampled voltage at a constant level, while the A/D converter translates it into its digital equivalent.

There are a number of ways to classify A/D converters. For instance, we could say that they are either "open-loop" or "closed-loop." Consider the typical analog-to-digital converter block diagram of Figure 8-1:

1. *Closed-loop*: If the dashed, digital feedback line were included, we would have a closed-loop A/D converter. In this case, a decision would be made by the decision circuits, and the result would be encoded into a digital format. The digital output would then be converted to analog by the D/A converter and compared to the original analog input. Should the two compare, the conver-

sion would be considered correct, and the conversion process would be terminated. If not, the process would continue.

2. *Open-loop*: In an open-loop system, the conversion is made and the digital result presented at the converter output for immediate use. The dashed digital feedback line would be eliminated from the diagram, and a validation comparison would not be made.

"Programmed" and "nonprogrammed" constitute another pair of terms used to describe A/D converters:

1. *Programmed*: The programmed A/D converter operates with a fixed, clocked sequence of events. When a conversion command is supplied to the A/D converter, it converts the input analog signal to digital format in a precise, prescribed sequence. All conversions are slaved to the externally supplied convert command, and they are not initiated by a change in input signal value.
2. *Nonprogrammed*: The nonprogrammed converter, on the other hand, is influenced by changes in the analog input signal. The converter bides its time until a signal change occurs. It then effects a new conversion so as to track the input. This type of A/D converter is often referred to as being *adaptive*.

One final way we will consider for the classification of A/D converters is to determine whether they utilize "discrete voltage comparison" or "capacitor charging."

1. *Discrete voltage comparison*: This sort of converter compares the analog input voltage to discrete reference levels to directly produce a digital output.

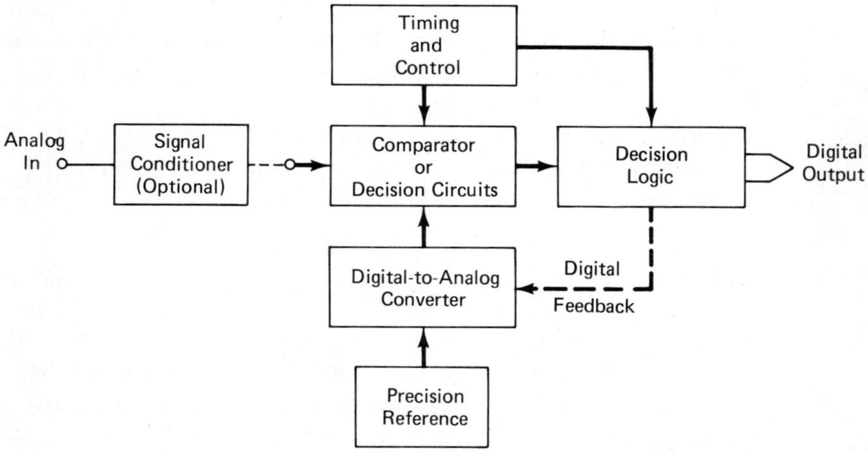

Figure 8-1. Typical Analog-to-Digital Converter Block Diagram

2. *Capacitor charging*: This sort of converter digitally encodes the time it takes to charge or to discharge a capacitor to the input voltage value, always with respect to some reference.

Each A/D converter type fits one of the categories in each classification. For instance, the typical successive approximation converter is closed-loop and programmed, and it utilizes discrete signal comparison.

8-1 THE CONVERSION PROCESS

The conversion process is much like making weight measurements with a balance scale (Figure 8-2a). Like the scale, the converter's digital output is only as granular as the smallest measurement increment. In this context, the smallest weight is equivalent to a quantum interval (Chapter 3). As we shall see later, there are many ways of supplying the balance "weights." They may be added one at a time and accepted or rejected. Or, in the other extreme, the decisions may be made on all balance "weights" simultaneously. Regardless of the method, the analog-to-digital converter output is the ratio of the input signal value with respect to the effective converter reference voltage.

Consider the basic converter of Figure 8-2b. In this diagram:

1. V_X = the unknown analog signal to be measured, and
2. X = the digital result, after measurement.

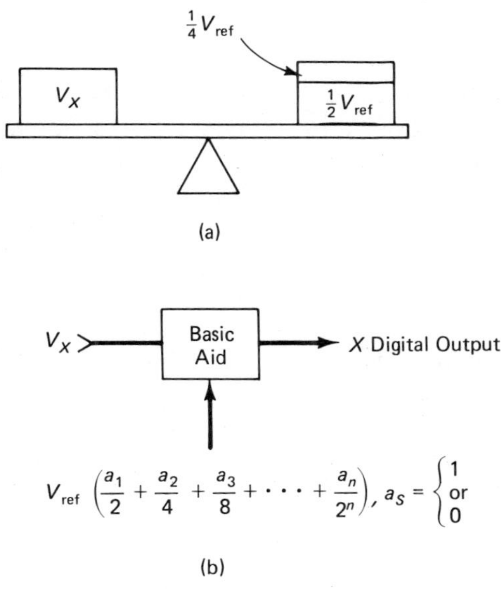

Figure 8-2. The Conversion Process

The third input:

$$V_{\text{ref}}\left[\frac{a_1}{2}+\frac{a_2}{4}+\frac{a_3}{8}+\ldots+\frac{a_n}{2^n}\right]=V_{\text{ref}}\sum_{n=1}^{n}\frac{a_k}{2^n}$$

where the a's are either a one or a zero, is a fraction of the converter's reference voltage, V_{ref}. You should also recognize that:

$$X = \text{digital output} = \frac{a_1}{2}+\frac{a_2}{4}+\frac{a_3}{8}+\ldots+\frac{a_n}{2^n}$$

When the conversion is complete and the circuit at balance:

$$V_X = V_{\text{ref}}\left[\frac{a_1}{2}+\frac{a_2}{4}+\frac{a_3}{8}+\ldots+\frac{a_n}{2^n}\right] \qquad (8\text{–}1)$$

or:

$$V_X = V_{\text{ref}}\,X \qquad (8\text{–}2)$$

and therefore:

$$X = \frac{V_X}{V_{\text{ref}}} \qquad (8\text{–}3)$$

Clearly, the digital output, X, does represent the ratio of V_X to V_{ref}. To take an example, consider an A/D converter in which:

$$V_{\text{ref}} = 10{,}000 \text{ V}$$
$$V_{FS} = 10{,}000 \text{ V}$$
$$n = 10 \text{ bits}$$

Let's compute the binary number, X, which is equivalent to the converter's highest output reading, that is, all 10 bits are one's. First calculate the value of:

$$V_{\text{ref}}\sum_{n=1}^{n}\frac{a_k}{2^n}$$

where the a's are all one's as shown in the following table:

Bit	*Binary Weight*	*Fraction of Reference*	*Actual Voltage*
a_1	1	$\frac{1}{1{,}024} V_{ref}$	0.0098
a_2	2	$\frac{1}{512} V_{ref}$	0.0195
a_3	4	$\frac{1}{256} V_{ref}$	0.0390
a_4	8	$\frac{1}{128} V_{ref}$	0.0781
a_5	16	$\frac{1}{64} V_{ref}$	0.1563
a_6	32	$\frac{1}{32} V_{ref}$	0.3125
a_7	64	$\frac{1}{16} V_{ref}$	0.6250
a_8	128	$\frac{1}{8} V_{ref}$	1.2500
a_9	256	$\frac{1}{4} V_{ref}$	2.5000
a_{10}	512	$\frac{1}{2} V_{ref}$	5.0000
Totals	1,023		9.9902 V

As we can see, the all-one's condition is equivalent to one quantum level less than the effective full-scale input analog signal value, that is,

$$\underbrace{9.9902\text{ V}}_{\text{All one's}} + \underbrace{0.0098\text{ V}}_{\text{LSB}} = 10.000\text{ V}$$

This result correlates directly with the discussion on quantization found in Chapter 3. There we established that a quantum interval (binary) was:

$$\text{Quantum interval} = \frac{V_{FS}}{2^n}$$

and that the highest quantizer output code would indeed vary from the theoretical maximum by one quantum level.

Using the 10-bit A/D converter of the previous example, consider a situation where $V_X = 6.29$. If expression (8-3) were used, we would obtain the following result:

$$X = \frac{6.29}{10.000} = 0.629$$

Unfortunately, this is merely a ratio and does not provide the actual binary output code being sought. If we recognize that $X_{\max} = 2^n$, then we can obtain any equivalent output code, x, by multiplying the ratio V_X/V_{ref} by 2^n, and we now have:

$$x = \frac{2^n V_X}{V_{\text{ref}}} \tag{8-4}$$

Proceeding with the example, we find:

$$x = \frac{1{,}024 \times 6.29}{10.000} = 644.096$$

The converter is, of course, accurate only to within $\pm\frac{1}{2}$ a quantum interval. So the fraction, 0.096, is discounted, and the digital output is binary 644. The converter's output digital code, x, in binary representation is found in the following table:

Bit	*Bit Output*	*Output Weight*
a_{10}	1	512
a_9	0	0
a_8	1	128
a_7	0	0
a_6	0	0
a_5	0	0
a_4	0	0
a_3	1	4
a_2	0	0
a_1	0	0
Total output weight =		644

For this A/D converter, 6.29 V is represented by a binary output of 644. Suppose the converter's internal reference and its full-scale input voltage were both 10.240 V. Once again, using expression (8–4), we find that:

$$X = \frac{1024 \times 6.29}{10.240} = 629$$

This is a very convenient situation, because the digital output representation correlates directly with the analog input voltage. Few manufacturers use a binary reference, however, and typical input ranges are 0 to 10 V, ±10 V, 0 to 5 V, ±5 V, and ±2.5 V.

If the code required were binary-coded decimal (BCD), rather than binary, the following expression would be used in place of expression (8–4):

$$x = \frac{10^m V_X}{V_{\text{ref}}} \qquad (8\text{–}5)$$

where m = the number of decades.

8–2 THE SUCCESSIVE APPROXIMATION A/D CONVERTER

The most popular A/D converter technique is that of successive approximation. So it is not surprising that its block diagram (Figure 8–3) looks very much like that of the typical A/D converter of Figure

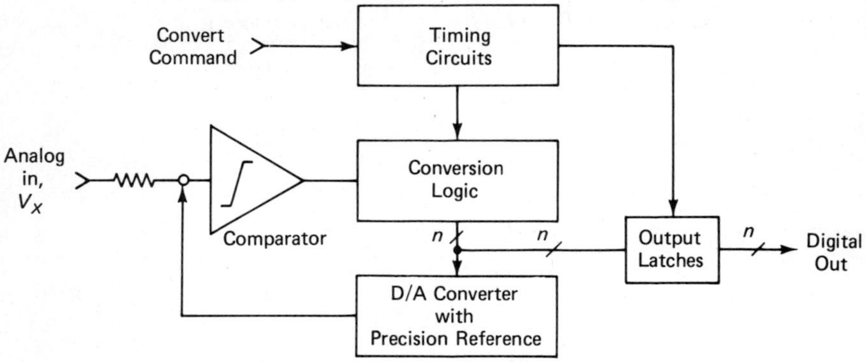

Figure 8-3. Successive Approximation A/D Converter

8-1. In terms of the previously defined categories, it is closed-loop and programmed, and it makes discrete voltage comparisons.

As the name "successive approximation" implies, it works on a "cut-and-try" method where each bit is tested against the analog input and either accepted or rejected. In this method, the heaviest weights (most significant bits) are tried first, and then the successively lower weights (least significant bits) are tested. Look at the weight-balance example of Figure 8-4a. First, the half-full-scale weight (most significant bit) was applied, found not sufficient to tip the balance, and, because more weight was required, was accepted. Next, the quarter-full-scale weight was applied. Once again the balance did not tip, and the weight was accepted. Finally, the eighth-full-scale weight was applied on top of the previous two. The balance was tipped, and the eighth-full-scale weight was rejected. If the converter had only three output bits, as is implied here, the conversion process would be terminated.

Looking at the result, we can say that the analog input falls somewhere between six-eighths and seven-eighths of full scale. Actually, if we consider the transfer function of the quantizer (Figure 3-1), we see that any given output code straddles the input level it represents by $\pm\frac{1}{2}$ a quantum level, and we know that the digital output is correct within one-half an LSB. A 3-bit decision tree, which depicts the successive approximation algorithm, can be seen in Figure 8-4b. As we see, first the quarter-full-scale decision is made, then the three-quarter- or one-quarter-full-scale decision. Finally the decision is on which eighth of the scale contains the answer. If more than three bits were involved, the tree would continue to expand to the right, in the manner shown in the figure.

Figures 8-5a and 8-5b show the operation of an actual 3-bit successive approximation A/D converter. The entire process is keyed to and synchronized with the external convert command. At the end of

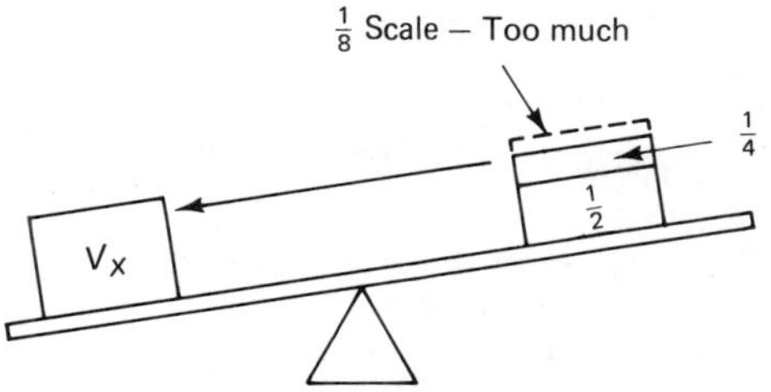

(a)

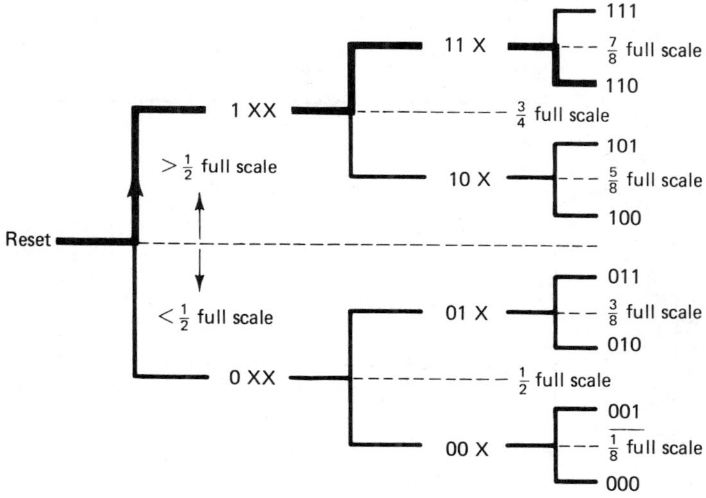

(b)

Figure 8-4. Conversion Process for 3-Bit Successive Approximation A/D Converter

the command pulse, the flip-flop FFA (MSB) is set, and FFB (MSB/2) and FFC (LSB) are reset. The D/A converter translates the binary number 100 to a half-full-scale analog voltage, which is compared to the analog input signal by the decision circuit.

The J-K flip-flops (FFA, FFB, FFC) operate in the following manner. If both the J and K inputs are high when a clock input (C) is applied, they change state. If both inputs are low, the state remains as it is. So had the half-scale voltage from the D/A converter been greater than the input analog signal, the decision circuit output would have gone high, the flip-flop J and K inputs would have been high, and the trailing edge of the T1 output would have changed the FFA state, thus removing the bit. For the case shown in Figure 8-5a, the D/A converter output was less than the analog input. This kept the decision cir-

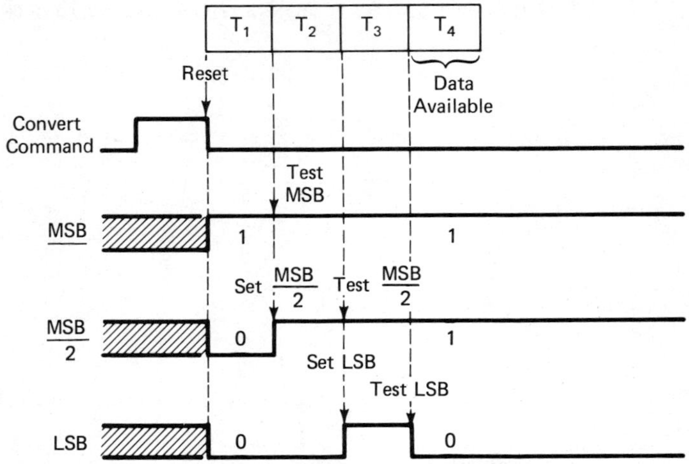

(a)

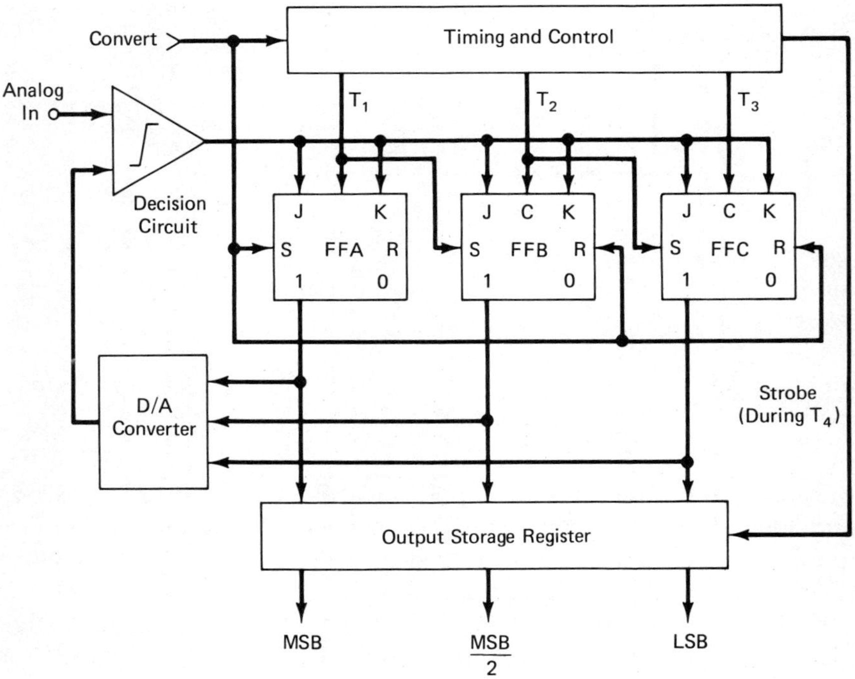

(b)

Figure 8-5. Successive Approximation A/D Converter Timing cuit output low so that the bit was not removed. At the same time that the T1 output tested the most significant bit (FFA), it also set the quarter-scale bit (FFB). The same decision process is then carried out during the T2 period. In this case, because the half-scale bit was retained, three-quarter-scale (1/2 + 1/4) is compared to the analog input. As Figure 8–5a shows, the input was greater than three-quarter-scale, FFB was not reset at the end of T2 and the one-quarter-scale bit also remains. The next, and final, decision is the seven-eighths-scale decision (Figure 8–4b), as the condition 11 X exists. When the end of T2 tested FFB, it also set FFC. The decision process is now repeated a third time, and, as Figure 8–5a indicates, the bit was removed. The final result is 110.

Had the converter contained more than three bits, this process described would have continued from MSB to LSB until all bits had been decided. The successive approximation A/D converter is often called the "bit-at-a-time" converter, and the time required for conversion is therefore defined by multiplying the number of bits by the time of a *T*-period. As our example shows, however, one time period is usually allowed to strobe the data into the output storage register and:

$$T_{\text{conv}} = (n + 1)\ T\text{-period} \qquad (8\text{–}6)$$

The *T*-period is determined by a critical timing loop that begins with the setting of any one of the bit flip-flops and ends when the decision circuit output has settled at the input to the same flip-flop. In the worst case:

1. the flip-flop must go from the reset to set condition;
2. the D/A converter output must settle;
3. the decision circuit must switch; and
4. its output must settle through the J and K inputs of the flip-flop.

For a slow-speed unit, the converter's accuracy is determined by the D/A converter (with its internal precision reference) and the decision circuit. For a high-speed unit, the various circuit time constants, along with their inherent settling time and internal converter noise, must be considered. Typical operating ranges for successive approximation converters are from 1 μsec conversion time with 8 to 10 bits, to 10 μsec with 12 to 14 bits. Unfortunately, aperture time for this converter type is essentially the same as the conversion time. To alleviate this situation, a sample-and-hold circuit is commonly utilized as a front-end signal conditioner.

Because the successive approximation converter requires fast operation of the critical timing loop, it also requires a wide-band decision circuit or comparator. This makes the instrument susceptible to errors caused by noise spikes that occur on the input. Such noise can be particularly troublesome at the decision points for the higher-order bits. Take the following example for a 10-bit converter:

	MSB ↓									LSB ↓	
Correct output	1	1	0	1	0	0	1	1	1	1	= 847
Actual output	1	1	(1)	0	0	0	0	0	0	0	= 896
			↑ Noise spike occurs								

A noise spike has occurred at the point where the decision is made to leave the eighth-scale weight or to remove it. Once the wrong decision is made, the decision circuit is forced to a permanently high position. Testing of the remaining seven lower-order bits now becomes academic. With the decision circuit locked high, their subsequent removal has already been determined. Once again, use of a sample-and-hold circuit can solve the problem. The S/H provides this narrow sample aperture and holds the value constant during the conversion period.

8-3 THE DUAL-SLOPE INTEGRATING A/D CONVERTER

The dual-slope integrating analog-to-digital converter (Figure 8-6a) is the most popular of the capacitor charging converters. It is also closed-loop and programmed. Operation of this converter can best be understood by following the example found in Figure 8-6b. When the convert command is applied, the one-shot multivibrator (OS) is triggered, and switch S1 is closed (switches S2 and S3 are left open). S1 shorts the integrating capacitor, C, bringing the integrator output A back to zero. The output from OS must be long enough to accommodate the situation where the integrator output had drifted off to full scale. On the falling edge of the OS output pulse, S1 is opened, and flip-flop FF1 is set. This closes switch S2 and enables the C1 side of the and-or logic, which controls the clock input to the counter. While the counter is counting clock pulses, the integrator is integrating the input analog signal, $+V_X$, in accordance with the expression:

$$-\frac{1}{RC}\int_0^{T_1} V_X\,dt \qquad (8\text{–}7)$$

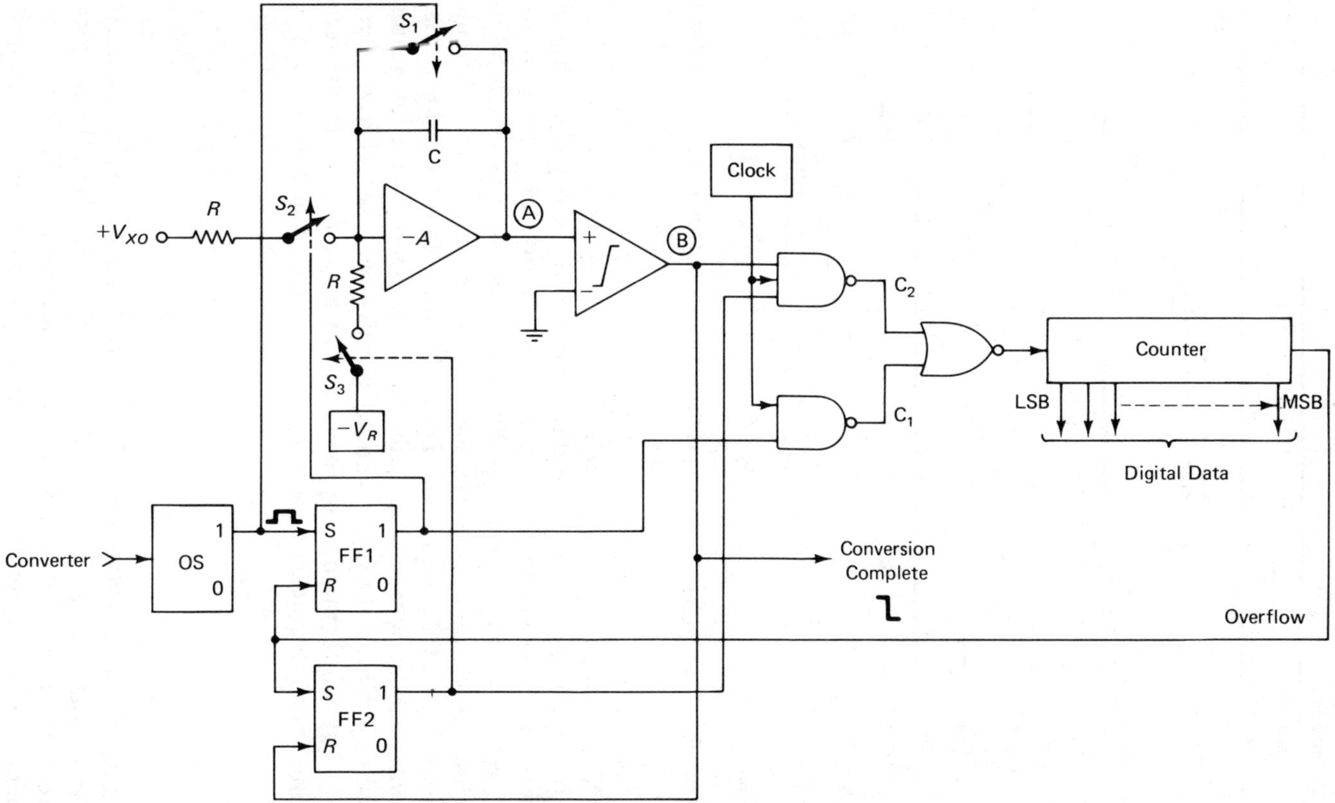

Figure 8–6a. Dual-Slope Integrating A/D Converter Block Diagram

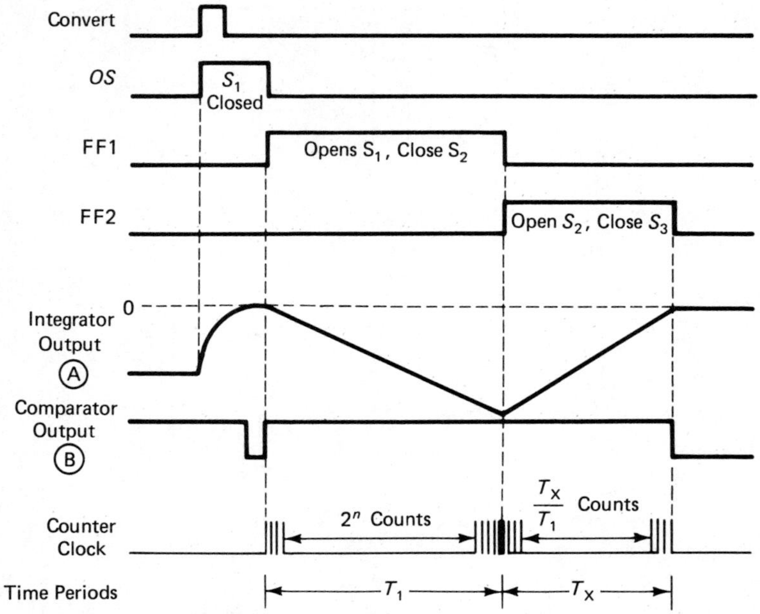

Figure 8-6b. Dual-Slope Integrating A/D Converter Timing Diagram

Note: *Output A goes negative.*

It continues this process until the counter fills or reaches its maximum count (time period T1). On the next count, the counter overflows, resetting FF1 and setting FF2. This action causes switch S2 to open and S3 to close. It also disables the C1 side of the and-or circuit and enables the C2 side. (Note: The third input to the C2 side of the gate was enabled by the comparator output B as soon as the integrator started to integrate the input signal.) In this state, the negative reference ($-V_R$) is integrated until the output A again returns to the zero baseline. The integration time is T_X, and the expression involved is:

$$-\frac{1}{RC}\int_0^{T_X} -V_R \, dt \qquad (8\text{–}8)$$

Integration continues until the integrator output reaches zero. At that time, the comparator output B switches, shutting off the clock and resetting FF2. The conversion process is complete. All during the T_X period, the counter has been counting. When the conversion is complete, the number contained represents the analog input that has been measured. As the negative reference ($-V_R$) is equal in magnitude to V_{FS}, a full-scale input would find the counter at its maximum count when the conversion was completed. (Actually, the maximum output is V_{FS} minus one count or quantum level.) Any lesser input voltage would provide a code that would be in direct proportion to its value, with respect to V_{FS}.

A very important and unique feature of the integrating A/D converter is that accuracy is independent of R, C, and the integrator offsets.

Considering V_X to be a dc value, this can be proven as follows:

During $\mathbf{T_1}$:

$$\text{Integrator output} = V_1 = -\frac{1}{RC}\int_0^{T_1} V_X\,dt$$

and where V_X is a constant (dc):

$$V_1 = -\frac{V_X}{RC}T_1$$

During $\mathbf{T_X}$:

$$\text{Integrator output} = 0 = V_1 - \frac{1}{RC}\int_0^{T_X} -V_R\,dt$$

and where $-V_R$ is also a constant:

$$V_1 + \frac{V_R}{RC}T_X = 0$$

Substituting for V_1:

$$-\frac{V_X}{RC}T_1 + \frac{V_R}{RC}T_X = 0$$

or:

$$-V_X T_1 + V_R T_X = 0 \qquad (8\text{-}9)$$

Considering that:

$$T_1 = 2^n\ t_{\text{clk}}$$
$$T_X = Xt_{\text{clk}}$$
$$t_{\text{clk}} = \text{clock period}$$
$$X = \text{final counter value after } T_X$$
$$2^n = \text{value of the full counter}$$
$$V_R X = V_X 2^n$$

and:

$$X = \frac{V_X}{V_R}\,2^n \qquad (8\text{–}10)$$

So, not surprisingly, we see that the converter output is a ratio of V_X with respect to V_R (see expression 8–4). We also see that R and C are eliminated at the step that produces equation (8–9). It is important to remember, however, that the converter accuracy still depends on the accuracy of the precision voltage reference ($-V_R$).

Because this type of converter actually performs integration on the input signal, it is not uncommon for the T1 integration period to be set equal to the period of an unwanted interference signal. The contribution due to the unwanted signal is greatly attenuated by the basic operation of the converter. Noise rejection that exceeds 60 dB is not uncommon.

Conversion time is determined by:

$$T_{\text{conv}} = (2^n + X)t_{\text{clk}} + T_D$$

where T_D = time to zero the integrator

Typical dual-slope converters range from 8 to 14 bits and 200 μsec down to about 200 msec. Although the examples provided have been based on a binary output, 3- and 4-digit BCD are very common. Many digital panel meters (DPM) use the dual-slope integrating technique.

8–4 THE SINGLE-SLOPE INTEGRATING A/D CONVERTER

Another type of integrating A/D converter is the single-slope design found in Figure 8–7a. This concept would be classified as open-loop, programmed, and capacitor charging. Unlike the dual-slope converter, the single-slope unit does not integrate the input signal. Instead, it compares the input with the output of an integrator, which is driven by the precision reference, $-V_R$.

To understand the operation of the single-slope circuit, consider both Figures 8-7a and 8-7b. Application of a convert command triggers the one-shot multivibrator to close switch S1. The one-shot pulse

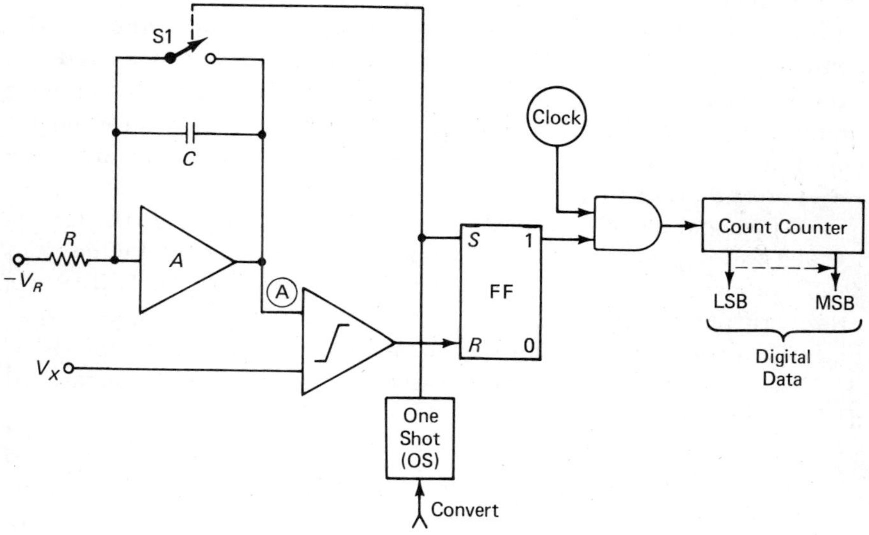

(a) Block diagram

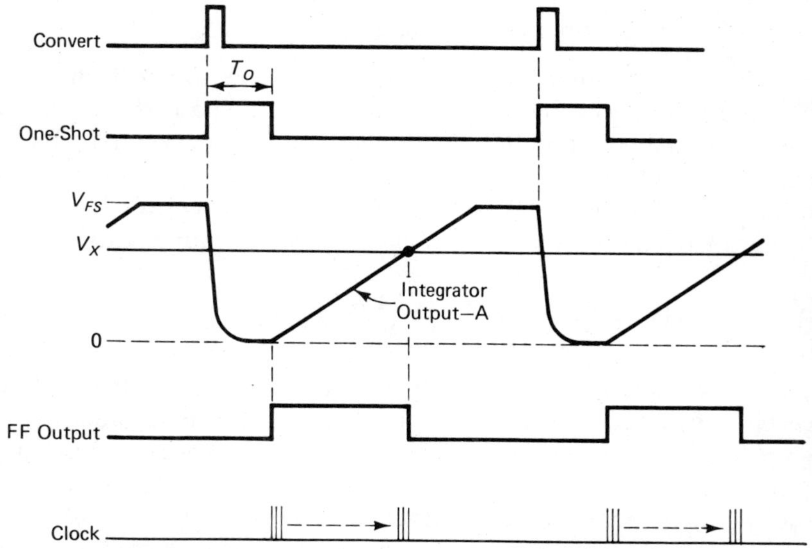

(b) Timing diagram

Figure 8-7. Single-Slope Integrating A/D Converter

width must be sufficient to return the integrator output A to zero. On the trailing edge of the pulse, switch S1 is opened and the count flip-flop (FF) is set. Simultaneously, the integrator starts to operate on the reference voltage ($-V_R$), and the counter starts to count clock pulses. When the integrator output exceeds V_X by an amount that is greater than zero but less than $\frac{1}{2}$ LSB, the comparator switches and FF is reset. This shuts off the clock and the counter is stopped. The number in the counter is equivalent to the value of the voltage being measured. As expected, a full counter (all one's) represents V_{FS} minus one count or quantum interval. Regardless of the value of V_X, the integrator output continues until its most positive level is reached and remains (driven by $-V_R$) until a new convert command is applied.

The accuracy of this converter is heavily influenced by the linearity of the integrator ramp. By trimming, accuracies of better than ±0.05 percent have been achieved at room temperature. Difficulties arise with temperature changes, since the RC components, as well as both the integrator amplifier and the comparator, are subject to drift.

Conversion time depends directly on the voltage being measured and is expressed by:

$$T_{\text{conv}} = t_{\text{clk}}X + T_D \qquad (8\text{–}11)$$

where t_{clk} = clock period
X = counter code equivalent to V_X
T_D = time-to-zero integrator

Typical commercial single-slope converters range from 8 to 14 bits and from 1 to 2 kHz down to 200 Hz. In the past, this type of converter has been used in digital voltmeters. It is discussed here due to its extreme simplicity for low-accuracy, low-speed applications. A 4- to 6-bit unit, which would operate at a few-kilohertz conversion rate, can be constructed out of a handful of inexpensive components.

8–5 THE SERVO A/D CONVERTER

The servo A/D converter, sometimes called the tracking A/D converter, can be seen in Figure 8–8a. It is a closed-loop, adaptive, discrete voltage measuring A/D converter. The best way to describe the converter's operation is to consider that V_X has been sitting at zero and then suddenly moves to a new level, as seen in Figure 8–8b. As soon as the input moves off zero, the comparator switches, placing the steering logic in the up position and enabling the count clock signal. The up/

down counter starts to count up. The counter output is D/A converted and compared with the input signal, V_X. When the counter value exceeds V_X, the comparator switches, stopping the count. The counter value is equivalent to the analog input within a count or quantum level.

Actually, the servo-type converter never stops converting. If the new level remains at a constant value, the counter output oscillates around the final level, as shown in the right-hand side of the diagram. When the first new level conversion is complete, the comparator

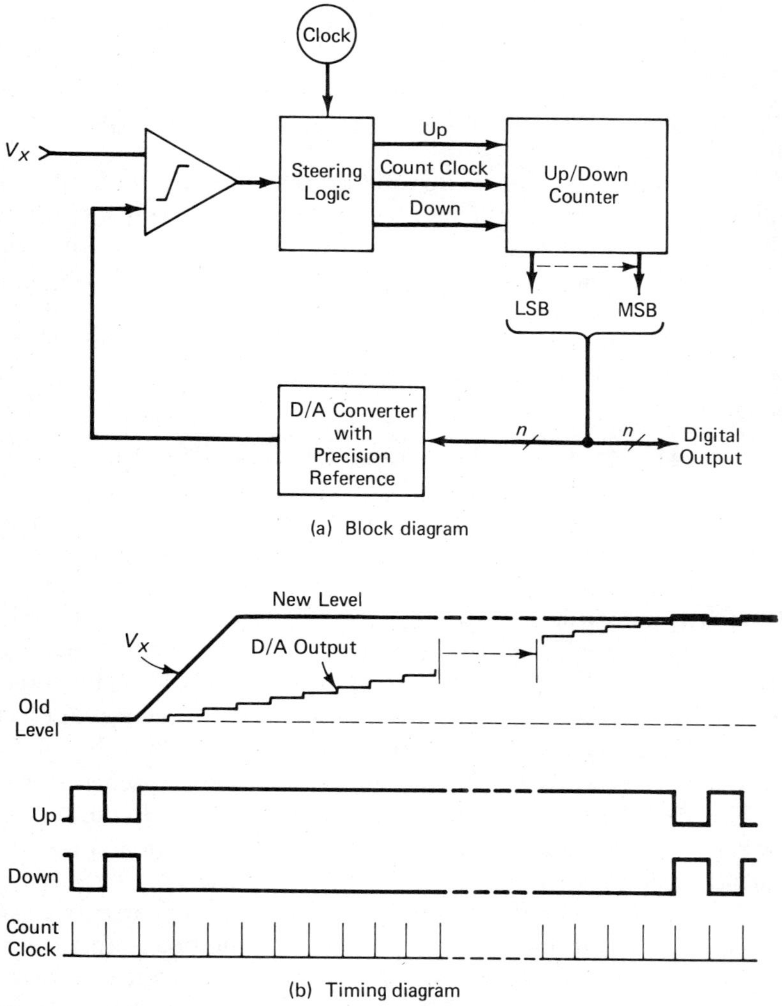

Figure 8-8. Servo A/D Converter

switching does not stop the counter. The steering logic simply switches from up to down, and the count clock starts again. As soon as the counter output is reduced by one count, the comparator switches again, the logic switches from down to up, and the next count clock adds the count back again. This action would continue until V_X changes again. The steering logic ensures that the counter is always commanded to count either up or down and never both at the same time.

This configuration, referred to as the servo A/D converter, acts like a typical follow-up servo. That is, when the input changes, the system reacts to reduce the error at the comparator input to zero. The converter is bidirectional and can track an input change in either direction. The system accuracy depends on the D/A converter voltage reference, the ladder network and switches, and the comparator gain and drift characteristics.

The conversion time is directly proportional to the input voltage change, V_X. A single-level change would occur within one clock pulse, while a full-scale change would require $2^n - 1$ clock pulses. As you would surmise, the servo A/D converter works very well where small input changes are the rule.

This converter type, often utilized for voice communications systems, is often referred to as a delta system.

8-6 THE RAMP A/D CONVERTER

The ramp A/D converter operation is very similar to that of the servo A/D converter, but it does not track or follow the input voltage. It is initialized by a convert command, makes one conversion, and stops. The converter, which can be seen in Figure 8-9a, is programmed, closed-loop, and discrete voltage measuring. When the convert command is applied (Figure 8-9b), the flip-flop (FF) is set. This resets the counter and enables the clock. The counter starts to count and its D/A converted output is fed back to the comparator. When the D/A converter output exceeds V_X, the comparator output switches and resets FF. This disables the clock and stops the counter. The counter output now represents V_X and remains at that value until a new convert command is received.

Unlike the servo A/D converter, which would adaptively cycle around a given output level if the input remained constant, the ramp A/D converter always starts from a reset position. That is, a new convert command would reset the counter, and it would have to count back up to the V_X equivalent count to complete the conversion.

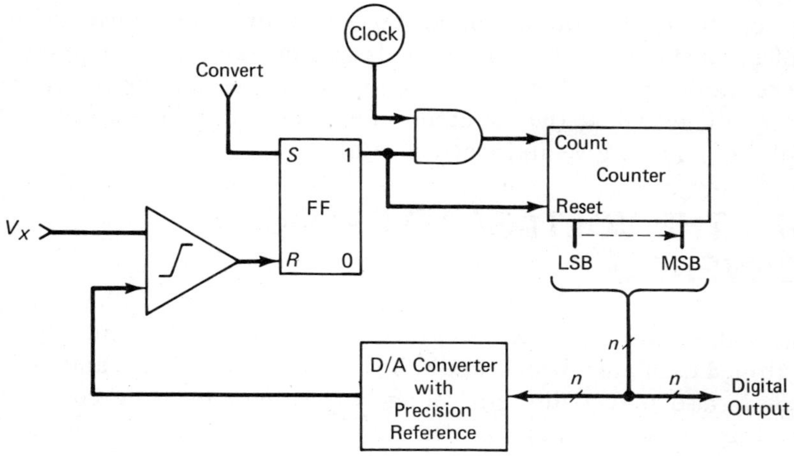

(a) Block diagram

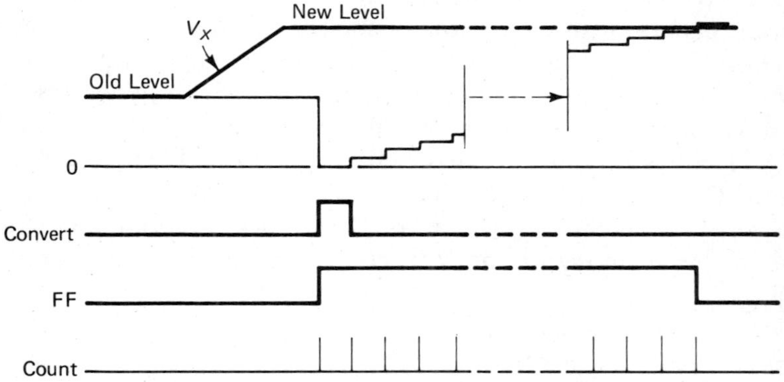

(b) Timing diagram

Figure 8-9. Ramp Type A/D Converter

The conversion time is totally dependent on the value of V_X and:

$$t_{\text{clk}(1)} < T_{\text{conv}} < t_{\text{clk}(2^n-1)} \qquad (8\text{–}12)$$

where $t_{\text{clk}(1)}$ = clock period

$2^n - 1$ = maximum counter value

Like the servo converter, the ramp converter's accuracy depends on the D/A converter system and the comparator characteristics. Accuracies could be expected to range from 6 to 10 bits, with conversion rates ranging from 4 kHz to 50 kHz. Like the single-slope integrating converter, the low-speed, low-accuracy units are inexpensive to implement. One note of caution: This type of converter can be susceptible to noise spikes. If a spike is wide enough to switch the comparator during a conversion, the counter could be shut off prematurely. This would, of course, give an incorrect result.

8-7 THE VOLTAGE-TO-FREQUENCY CONVERTER

The voltage-to-frequency (V/F) converter is a precision, voltage-controlled oscillator, which would be classified as adaptive and closed-loop. It also makes discrete voltage measurements. The converter's output is a series of square pulses whose frequency is equivalent to the voltage being measured.

The V/F converter block diagram can be seen in Figure 8-10. When a voltage (V_X) is applied to the converter, it causes the precision current generator to charge the capacitor, C. When the voltage on C reaches a specific value, the Schmitt trigger changes state. This state change activates the one-shot multivibrator, which generates a standard width pulse. The one-shot pulse, which after buffering becomes the V/F converter output, also discharges C through a precision charge dump. Once C has been discharged, the whole sequence is repeated. The result is a chain of pulses whose frequency is representative of V_X. Key to the accuracy of the V/F converter are the precision current

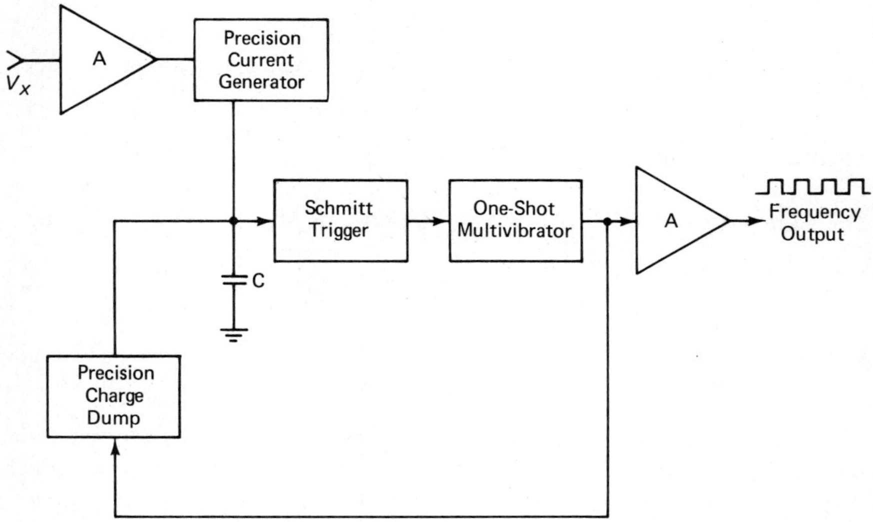

Figure 8-10. Voltage-to-Frequency Converter

source and the precision charge dump. To obtain good linearity, the output of the current generator must precisely track the value of the V_X, the voltage being measured. The precision charge dump must always return the capacitor voltage to the same precise discharged value.

The output pulse train from the V/F converter is usually applied to a digital counter. If the counter is cleared or set to zero and read on a periodic basis, its content truly represents frequency, or pulses per unit time. Proper selection of the counter accumulation cycle allows the counter contents to represent the converter input voltage. Conversion time is equivalent to the accumulation period. The converter is adaptive in nature, because it is always adjusting to keep its output frequency tracking the input voltage.

Very accurate V/F converters are available with typical specifications of:

Nonlinearity:	±0.005% of FS + ±0.005% reading
Temperature coefficient:	15 to 20 ppm/°C @ full scale
Full-scale frequency:	10 to 100 kHz

8-8 THE PARALLEL OR FLASH A/D CONVERTER

Perhaps the highest-speed A/D conversion technique available is the parallel or flash A/D converter, which can be seen in Figure 8-11. This converter implementation utilizes a decision circuit or comparator for each quantum interval within the quantization range. If we look carefully at the quantizer transfer function of Figure 3-1, Chapter 3, we note that there is one less quantum *interval* than there are quantum levels. So if our converter range consisted of 2^n levels, we would expect to have $2^n - 1$ intervals. Likewise, the parallel A/D converter has $2^n - 1$ comparators. Each of the comparators is biased to change state at the transition point for the output code it represents, as shown in Figure 3-1. That is, comparator #1 switches at the point depicted for level 1 in the figure; comparator #2 switches at the level 2 point; and so on. A depiction of the operation of a 2-bit converter of this type can be seen on the left-hand side of Figure 8-12b.

The power of this converter technique is that all comparators examine the input voltage simultaneously (in parallel) and make an immediate, total conversion. The comparator outputs are not directly usable, however, since they do not represent any binary, Gray, or BCD

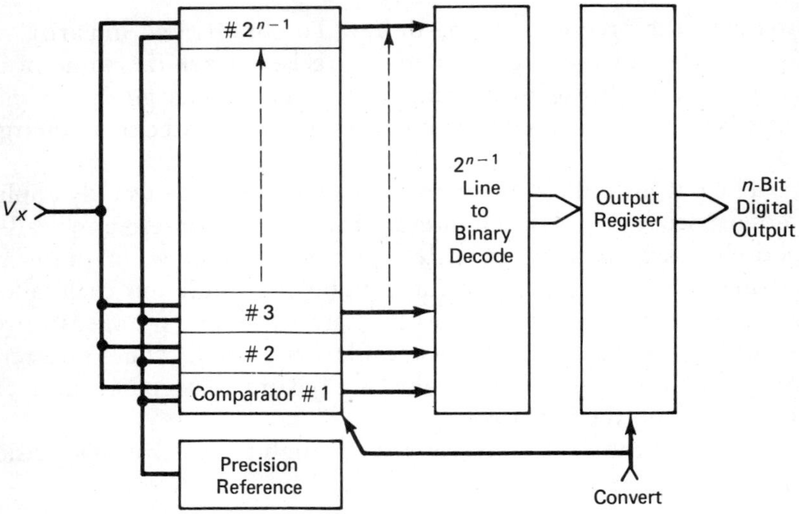

Figure 8-11. Parallel or Flash A/D Converter

code. Refer back to Figure 8-11: If the input value lies between $\frac{1}{2}$ and 1, comparator #1 is set on application of the convert command. If the input is between 2.5 and 3, comparators #1, #2, and #3 are all set. Should the input voltage lie between 9.5 and 10, all comparators from #1 through #10 are set. To use this unusual output, a $(2^n - 1)$ to binary decoder is used. The resulting binary output is set into the output register by the convert command.

The parallel, or flash, A/D converter makes discrete voltage measurements, works open-loop, and is programmed. When the convert command initiates a new conversion, it sets the results of the previous one into the output register.

The limiting factor for this type of converter is the number of comparators required for a relatively small number of bits. For instance, even a 6-bit unit would require 63 comparators. Seven- and 8-bit units (127 and 255 comparators) become totally unwieldy. Currently available units range from 4 bits at 50 to 75 MHz to 6 bits at 30 MHz.

8-9 THE PARALLEL, FEED-FORWARD A/D CONVERTER

The parallel, feed-forward A/D converter combines the concepts of parallel A/D conversion in a different architecture to reduce the number of comparators required for a given number of bits, while retaining a reasonably high conversion rate. The parallel, feed-forward A/D

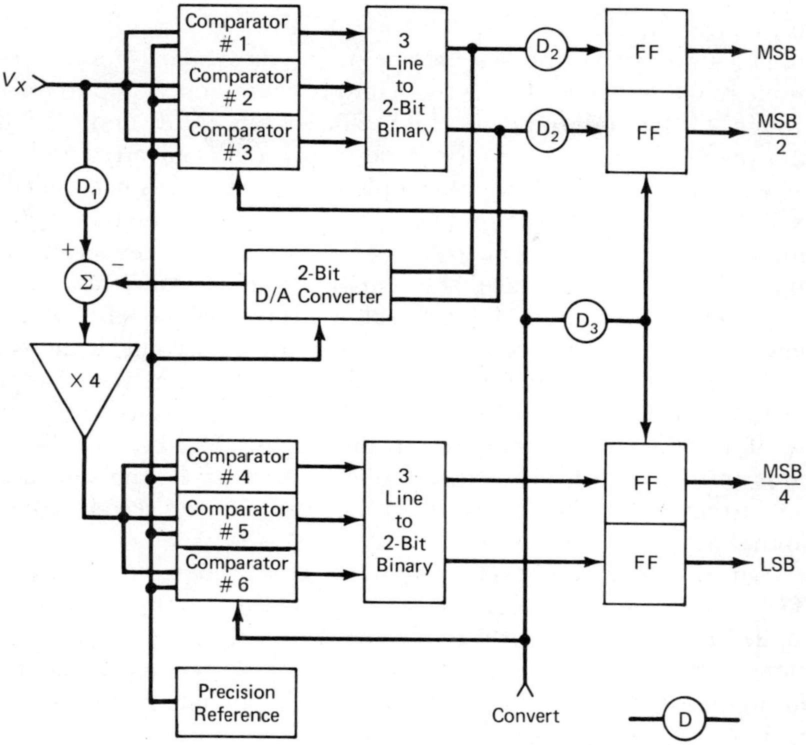

(a) Block diagram

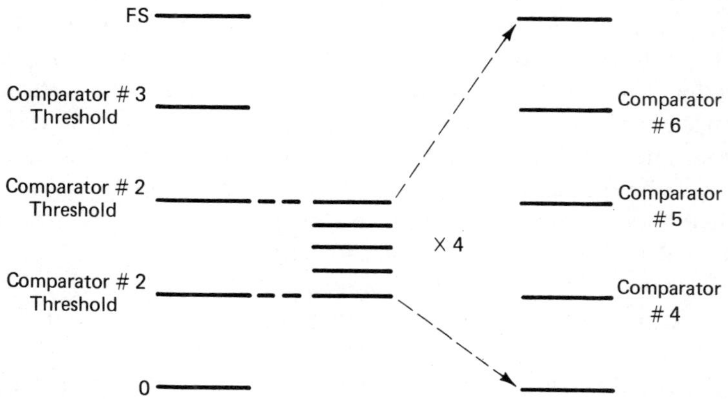

(b) Operation diagram

Figure 8–12. Parallel Feed-Forward A/D Converter

converter block diagram can be seen in Figure 8–12. If you look carefully at each of the two conversion groups and ignore the feed-forward hardware, it is easy to recognize that they are each 2-bit parallel or flash converters.

To understand the system operation, let's follow a conversion sample to completion. On the first convert command, comparators #1 through #3 determine the two higher-order bits (MSB and MSB/2). After the result has been binary encoded, it is D/A converted and subtracted from the original input sample point, which has been delayed (by D3) in order to arrive at precisely the same time. The remainder is amplified by four and applied to the second parallel converter. The multiplication by four places the remainder on the same scale as the original input, as seen in Figure 8–12b. This has two benefits. First, it allows the two parallel converters to be identical; second, it increases the effective value of a quantum interval at the second converter input. The first converter always determines where the input is with respect to any of the first three comparator thresholds. The combination of delays through the feed-forward circuitry must cause the subtracted result to arrive at lower converter input prior to the second convert command.

On the second convert command, the two lower order bits (MSB/4 and LSB) are decided. The results are binary encoded and supplied to the two output flip-flops. At the same time that these two bits are applied to their respective output flip-flops, the results of the two higher-order bits have emerged from the D2 delay units and are available to their respective flip-flops. All four are simultaneously strobed into the flip-flops by the convert command, delayed by D3. Note that, when the second convert command is completing the first conversion at the lower converter, it is also starting the second conversion with the upper converter. While two conversion periods are needed for the first full conversion output from the converter, new samples are available on each successive convert command that follows. The converter simply has an inherent "black box" delay of two conversion periods. While this arrangement sacrifices something in conversion rate with respect to a straight parallel converter, the feed-forward technique allows 4 bits to be determined with 6 comparators as opposed to 15 for the flash converter.

A number of these converters have been constructed. At one end of the range of models is an 8-bit, 5-MHz machine (Raytheon's AD–50), which utilizes 8 single-bit converter paths (8 comparators versus 255 for full parallel) with feed-forward between each. At the other end is a 6-bit, 50-MHz unit that has two 3-bit paths (14 comparators versus 63 for full parallel). More recently, the basic feed-forward concept has been utilized with other converter types, such as successive approximation, in place of the parallel converters. In general, these alternate architectures do not approach the speeds obtainable with the parallel feed-forward system. They can, however, make substantial speed improvements over that of the basic technique being used for the individual converter paths.

8-10 THE CYCLIC A/D CONVERTER

The cyclic A/D converter involves a totally different concept from any of those discussed in the preceding sections. As a converter, it would be classified as adaptive, open-loop, and discrete voltage measuring. It is highly analog in nature.

A typical converter stage can be found in Figure 8-13. Looking at Figure 8-13a, we see that it consists of n-identical converter stages, where n equals the number of bits. Conversion is accomplished on a bit-at-a-time basis, starting with the most significant bit. Each stage contains an offset network that centers the input about center scale, along with a decision circuit to determine whether the stage digital output should be a one or a zero. The absolute value circuit is, in fact, a full-wave rectifier, which folds the full-scale range about its midpoint. The absolute-value circuit is followed by a "×2" amplifier.

The best way to understand the converter's operation is to see what happens if a full-scale ramp (as seen at the top of Figure 8-13c) is applied to its input. Figure 8-13c shows the outputs of the first three stages of a 4-bit converter. At the first stage, the ramp is centered about 0 V and the bit decision made about that point. While the ramp is below 0 V, the bit output is a 0; and when above, the output is a 1. In the top part of conversion segment 2, we see that the top part of the ramp has been folded down by the stage 1 absolute value circuit and amplified by two by its ×2 circuit. The offset network of stage 2 has recentered the waveform about 0 V. The lower part of conversion segment 2 indicates the stage 2 decision circuit output.

Proceeding to conversion segment 3, we see that the upper waveform of segment 2 has been folded again by the stage 2 absolute value circuit and again amplified by two by the ×2 circuit. The stage 3 offset network rezeroes the waveform, and the stage 3 decision circuit output is as shown at the bottom of conversion segment 3. This operation continues for as many bits as the converter contains. If you look vertically at the stages' digital outputs of Figure 8-13c, you realize that the output is a type of Gray code. To complete the conversion, the Gray code is translated to binary and placed in an output buffer register.

The full-wave rectification at each stage also doubles the frequency. In an n-bit converter, the last stage (which does not need the absolute value and ×2 circuits) must operate at a frequency that is 2^{n-1} times the input frequency. For instance, for a 10-bit converter, whose input signal range was 100 kHz, the last stage would have to operate at 51.2 MHz.

Cyclic A/D converters ranging from 10-bits at a 10-MHz sample rate to 8 bits at a 15-MHz sample rate have been implemented.

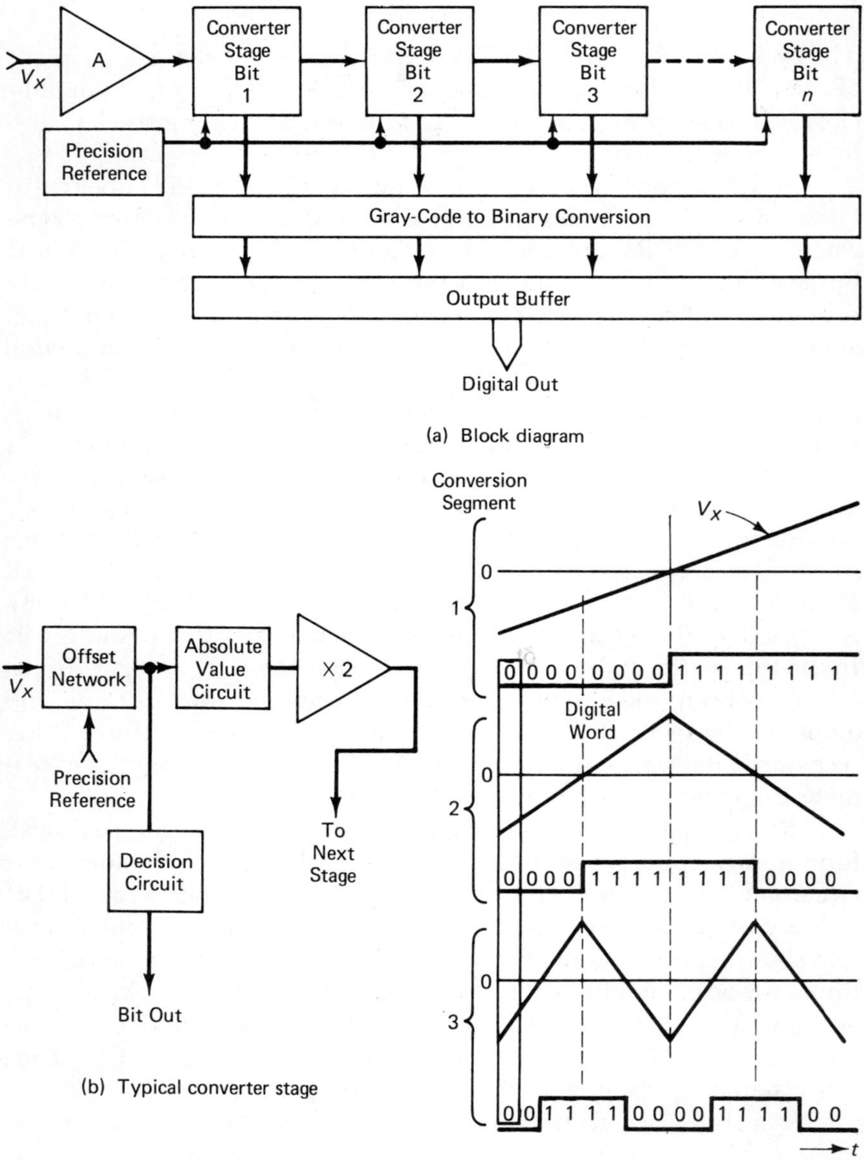

(a) Block diagram

(b) Typical converter stage

(c) Transfer functions–First three stages

Figure 8-13. Cyclic A/D Converter

PROBLEMS

8–1. A 12-bit A/D converter with an internal precision reference of +10,000 V is showing an output binary code of 101011001001. What is the input voltage V_X within $\pm\frac{1}{2}$ LSB accuracy?

8–2. A 3-digit A/D converter with an internal precision reference of +10,000 V has a voltage of 7.36 V applied to its input. What will the output code be?

8–3. The successive approximation A/D converter shown in Figure 8–5b has the following circuit delays:

FF()—Set to output:	10 nsec
—J-K input soak:	5 nsec
D/A converter delay:	50 nsec
Decision circuit delay:	20 nsec

What is the conversion time for a 10-bit operation?

8–4. A dual-slope integrating A/D converter has a resolution of 12 bits and an internal 100 kHz clock. Assume a capacitor dump time of 2 msec.

a. What is the minimum and maximum time?
b. What is the conversion time for an output code of 3068?

8–5. A 10-bit servo A/D converter and a 10-bit ramp A/D converter are both used to measure the same signal. Both have an internal 100 kHz clock. At the last previous sample, both instruments had an output code of 978. Presently, both have an output of 952. How long did it take each to complete the conversion?

8–6. Develop the logic circuits that must be used to implement a complete A/D converter with the V/F converter of Figure 8–10.

8–7. A data acquisition system contains an 8-bit parallel or flash converter.

a. How many comparators, or decision circuits, are needed in the converter?
b. If the flash converter were replaced by a two-section parallel feed-forward converter, how many would be needed?

8–8. Complete the diagram of Figure 8–13c for the 4-bit A/D, and write out the Gray code.

a. If the input frequency was 100 kHz, what frequency is applied to the last stage decision circuit?

REFERENCES

1. D. F. Hoeschele, *Analog-to-Digital/Digital-to-Analog Conversion Techniques*. New York: John Wiley & Sons, 1968.
2. H. Schmid, *Electronic Analog/Digital Conversions*. New York: Van Nostrand Reinhold Co., 1970.
3. D. H. Sheingold, *Analog–Digital Conversion Handbook*. Norwood, Mass.: Analog Devices, Inc., 1972.

9

System Error Analysis

9-0 INTRODUCTION

The accuracy of a data acquisition system is a function of all the various error sources within the system. These error sources generally confine themselves to the front end of the data acquisition system—or rather to those elements up through and including the analog-to-digital converter. In the preceding chapters, we have considered the inaccuracies or error contributions of each of these functions separately. In this chapter, we will discover how the individual error sources combine to effect the total system accuracy. We will also learn how to treat each of the contributing factors in the final system error calculation.

9-1 dc ERROR CIRCUIT MODELS

The block diagram of Figure 9-1 contains the dc error sources for the typical data acquisition system. When we considered each of the system functions independently, their peculiar error sources did not seem particularly formidable. Seeing them all together, as in Figure 9-1, we realize that the system situation can be very complex. For instance, instead of treating a simple offset voltage problem in the signal condi-

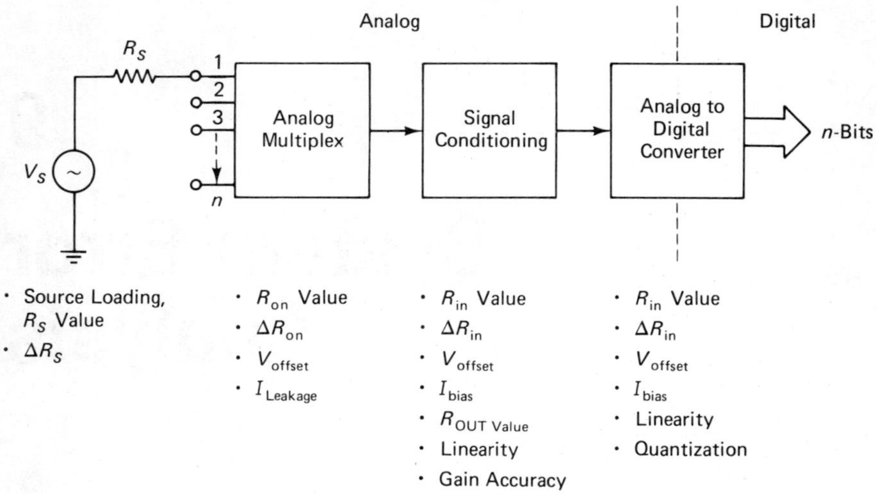

Figure 9-1. Data Acquisition System Block Diagram Showing dc Error Sources

tioner, we must take into account the front-end analog switch offset, as well as the input offset contributed by the A/D converter that follows.

To understand how the various parameters combine and interact, examine the diagrams of Figures 9-2, 9-3, and 9-4. As the signal conditioner invariably provides active isolation, the problem must be considered in three separate parts. These are:

1. the front end up through the signal conditioner input (Figure 9-2),
2. the signal conditioner (Figure 9-3), and
3. from the signal conditioner output through the A/D converter input (Figure 9-4).

First consider the circuit of Figure 9-2. The major path is a series dc circuit that starts at the source, continues through the "on" multiplexer switch and ends with the input network of the signal conditioner. Although polarities have been assumed for the various current and voltage sources, they may individually be of either polarity. In some cases, they add, and in some they subtract. The ac generators used for ΔV_{offset} depict the fact that these source elements are variable with such things as input voltage, power supply voltage variations, and temperature.

Figure 9-3 shows the gain error problem for both the inverting and noninverting operational amplifiers. As we see in the accompanying expressions, even the nominal gain is affected by the circuitry that

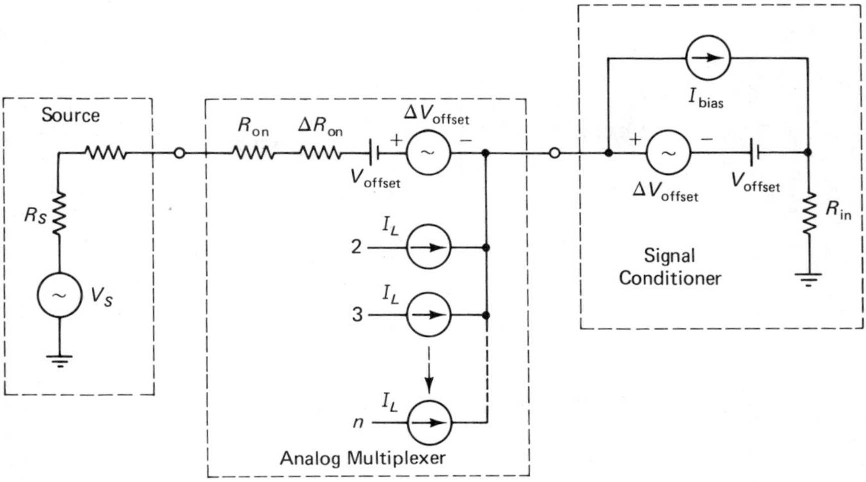

Figure 9-2. dc Errors: Source, Multiplexer, and Signal Conditioner Interface

precedes the signal conditioner. With reference to Figure 9-3a, the resistance, R_T, includes both R and R_{on}. We know from Chapter 6 that, due to the typically low input resistance, R_T has the greatest effect on the inverting amplifier. The models shown assume that the basic amplifier open-loop gain is very high. Although such is seldom the case, if A_{OL} suffered substantial reductions over the range of temperature and power supply voltages, it too would have to be considered. The data amplifier (Figure 6-17) would much resemble the noninverting amplifier of Figure 9-3b, but with two inputs.

The last dc circuit, seen in Figure 9-4, shows the interface between the signal conditioner and the A/D converter. Similar to the circuit of Figure 9-2, a series connection of resistances and generators start at the signal conditioner output and continue on through the A/D converter input. Again, generator polarities and directions have been assumed, and they may be either way. Also, an analog accuracy specification goes with every A/D converter. This specification is stated as: "$\pm x\%$ of full scale" or "$\pm x\%$ of full scale $\pm$ $y\%$ of reading." These factors cover the internal A/D converter nonlinearities, as well as the gain and offset shifts. Such parameters are as much a part of the accuracy calculation as the components found in the circuits of Figures 9-2 through 9-4. Last but not least, there is the quantization accuracy to be considered. The converter output is only correct when the input voltage is exactly equal to the digital code out. In the worst case, the output is inaccurate by $\pm\frac{1}{2}$ a quantum level or least significant bit.

Fortunately, an examination of the relative contributions of the

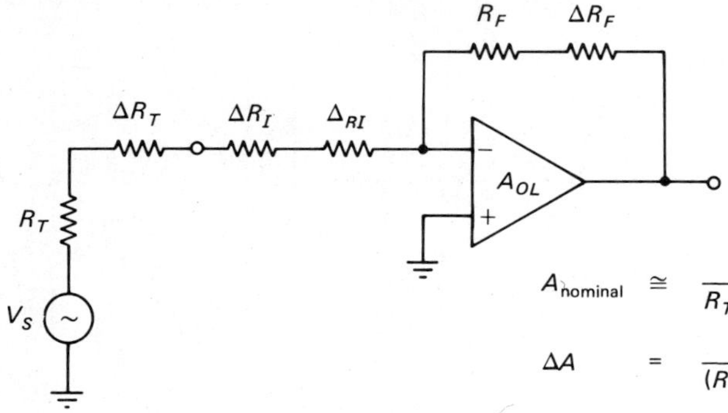

$$A_{\text{nominal}} \cong \frac{R_F}{R_T + R_I}$$

$$\Delta A = \frac{R_F + \Delta R_F}{(R_T + \Delta R_T) + (R_I + \Delta R_I)}$$

(a) Inverting operational amplifier

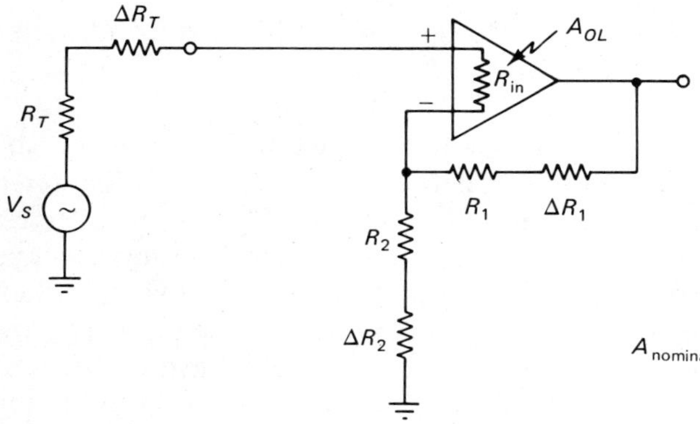

$$A_{\text{nominal}} \cong \left[\frac{R_{\text{in}}}{R_T + R_{\text{in}}}\right]\left[\frac{R_1 + R_2}{R_2}\right]$$

Input Divider — Normal Gain

$$\Delta A = \left[\frac{R_{\text{in}}}{(R_T + \Delta R_T) + R_{\text{in}}}\right]\left[\frac{(R_1 + \Delta R_1) + (R_2 + \Delta R_2)}{R_2 + \Delta R_2}\right]$$

Input Divider — Gain

(b) Non inverting operational amplifier

Figure 9-3. Typical Gain Error Problem

various errors shows that the equivalent circuits may be simplified for many data acquisition examples. For instance, multiplexer leakage current and offset voltage are not usually problems when electromechanical switches are used. Also, if the signal conditioner input impedance is very high, the source loading effects of R_S and R_{on} may be ignored. Such is particularly the case when reed relay multiplexers are used. In any case, circuit models should be derived for any proposed system; error source contribution should be analyzed, and the circuit model reduced to its most appropriate form.

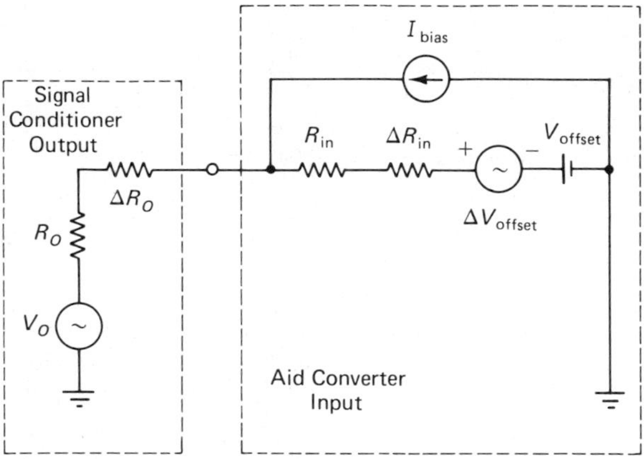

Figure 9-4. Typical dc Errors at the Signal Conditioner/Analog-to-Digital

9-2 ac ERROR CIRCUIT MODELS

Just as in the dc case, a number of ac or dynamic error sources are connected with the data acquisition system, as shown in the block diagram of Figure 9-5. As we see, ac response is a factor that appears under each functional system element. Indeed, the bandwidth and settling time of the system are functions of each of the individual element characteristics.

Figure 9-6 contains the settling time diagram for the typical data acquisition system. As in the dc case, the diagram is separated into three parts, with the amplifier providing isolation:

1. the source, multiplexer, amplifier interface,
2. the basic amplifier settling characteristics, and
3. the amplifier output, A/D converter input interface.

To establish the total system settling time, the individual settling times are combined as follows:

$$t_{\text{settle(total)}} = 1.1\sqrt{t^2_{\text{settle 1}} + t^2_{\text{settle 2}} + t^2_{\text{settle 3}}} \qquad (9\text{–}1)$$

Sample rate is also found associated with all the elements. If the data acquisition system consists only of an A/D converter, it determines the sample rate. Should an analog multiplexer be used, sample rate is established there. When the signal conditioner is a sample-and-hold, and when a multiplexer is not used, the sample-and-hold is responsible for the sampling rate. Aperture time is another parameter that may be determined by a different system function, depending on the system configuration. If a sample-and-hold is utilized, it determines

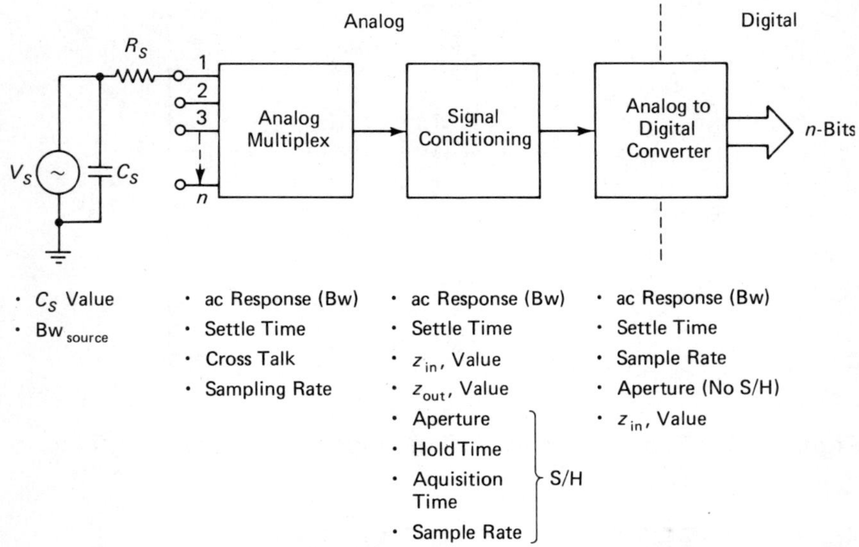

Figure 9–5. Data Acquisition System Block Diagram with ac Error Sources

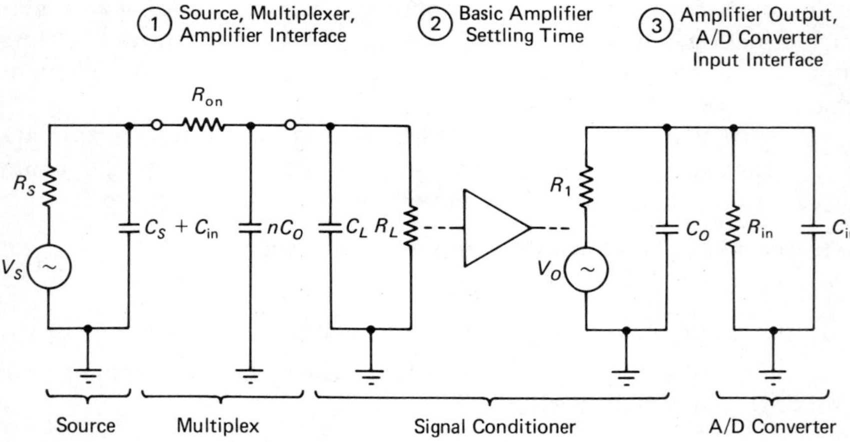

Figure 9–6. Settling Time Diagram

aperture. If not, aperture time is a function of the A/D converter.

As in the dc case, the designer must consider the relative impact of the ac parameters on the overall system accuracy. For example, if very narrow bandwidth signals are being measured, the importance of settling time and ac response may be negligible or even eliminated. Also, a very low source impedance can totally eliminate settling time as a factor in the analog multiplexer. Many other trade-offs of this nature can often be made.

Noise, both at the system input and as contributed by each of the system components, also leads to DAS error. Noise can cause instantaneous measurement or conversion errors. It can also be aliased during the sampling process, or it can establish a threshold below which measurement cannot be made. The designer must always keep it in mind.

9-3 ERROR BUDGETING

When designing a data acquisition system and establishing the required accuracy, you must give proper error allocations to each item of hardware that is to be included in the system. This procedure, where accurracies are assigned to each element, is called *error budgeting*. These error allocations should be properly weighted and achievable. For instance, A/D converters that convert to 10 bits of resolution in 500 nsec, and amplifiers that settle to 0.1 percent in less than 200 nsec are much easier to design than a 100-channel analog multiplexer that settles in less than 200 nsec. In this case, the multiplexer would be given a greater portion of the error budget. Error allocations should aim to make the design of each system element be of equal difficulty. They should be flexible and should allow for treatment of overlooked sources of error. The original error budget should serve as the design baseline.

As the design proceeds, constantly re-evaluate and modify the error budget. Some allocations will prove difficult to meet, while others will often be less of a problem than anticipated. Designers should be prepared to accommodate such budget changes to allow the best balanced design they can provide. Although simple circuit/system models can often be utilized in the final analysis, designers must understand the error mechanisms to the point where realistic adjustments and trade-offs can be made.

9-4 ERROR ANALYSIS

Historically, two basic methods have been utilized for data acquisition error analysis. They are the *worst-case method* and the *total-statistical approach*. These two techniques are almost diametrically opposite to one another.

The worst-case method assumes that all errors can occur at their maximum, worst-case value, in the same direction, at the same time. So the total worst-case system error is the algebraic sum of these worst-case values. As the error sources are affected by a multitude of independent factors, such an event is highly unlikely. The use of the worst-case method can render a system design impossible. At best, it can place such tight tolerances on the hardware elements that the design becomes much more costly and complex than is necessary.

The total-statistical approach assumes that all system parameters are statistically indepedent and vary in a random fashion with respect to one another. This technique views each parameter as following the normal distribution of Figure 9-7. In this figure, m is the mean or average value and:

σ = RMS value or standard deviation—68.3 percent of the time the actual error will be less than the σ value.

2σ = $2 \times$ RMS value—95.4 percent of the time the actual error will be less than the 2σ value.

3σ = $3 \times$ RMS value—99.7 percent of the time the actual error will be less than the 3σ value.

To illustrate, consider a field-effect multiplexer switch transistor, whose nominal or mean value of R_{on} is 100 Ω. Suppose also that long-term measurements for a large production run of these devices indicate a standard deviation from the mean (σ) of ±1 Ω. Let's not randomly select several thousand lots of 10,000 devices each. Then, on the average, from each lot:

1. 6,830 transistors would have R_{on} of 100 Ω ± 1 Ω or less,
2. 9,540 transistors would have R_{on} of 100 Ω ± 2 Ω or less, and
3. 9,970 transistors would have R_{on} of 100 Ω ± 3 Ω or less.

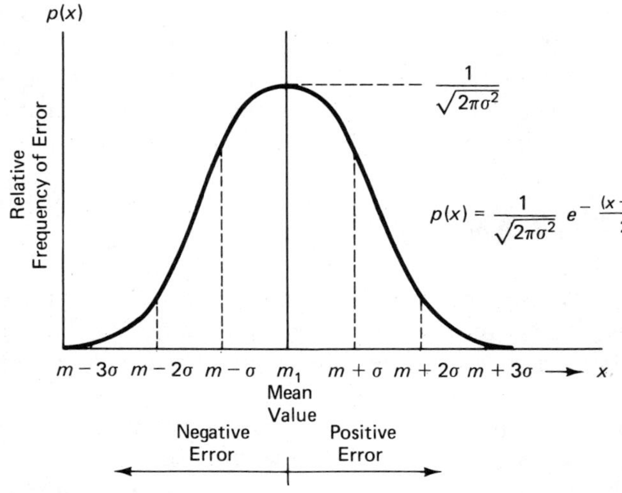

Figure 9-7. Normal Distribution

When building precision instruments or systems, it is not uncommon to select devices with tight tolerances. In the preceding situation, the transistors selected might be those within the $\pm\sigma$ limits ($100\ \Omega \pm 1\ \Omega$). This totally destroys the distribution since there is a hole in the middle. When buying devices or components, try to determine whether prior tolerance selection has taken place.

In the totally statistical error analysis approach, the root-sum-square (RSS) value of all the error sources is used, as defined by the following expression:

$$\text{Error}_{\text{total}} = \left[\sum_{n=1}^{n} \epsilon_I^2\right]^{1/2} \qquad (9\text{–}2)$$

where ϵ_I = the individual error sources

or:

$$\text{Error}_{\text{total}} = \sqrt{\epsilon_1^2 + \epsilon_2^2 + \epsilon_3^3 + \ldots \epsilon_n^2} \qquad (9\text{–}3)$$

The values used are usually either the 3σ values or the maximum value from the manufacturer's data sheet.

Unfortunately, not all error sources are independent from one another. It is not uncommon for several different error sources to be influenced in the same direction by a common factor. In opposition to the algebraic method, the total statistical approach could very well produce a system with insufficient overall accuracy.

How then should the problem of system accuracy be handled? Let's first categorize the types of error sources encountered in a typical data acquisition system. The three general classes of error sources are:

1. mean or average errors,
2. systematic errors, and
3. statistical or random errors.

Mean or Average Errors

A mean, or average error, is the center, or nominal value, from which a parameter varies. In the preceding multiplexer transistor example, $100\ \Omega$ was the average value. Often, average values of offset, leakage, bias, and the like can be adjusted out or compensated for in an operational system.

Systematic Errors (Parametrics)

Systematic errors are those which always vary as some definite function of the system operating conditions. An example would be a semiconductor multiplexer leakage current that doubles for every 7.5 to 10° C. Some of the causes for these variations are temperature, power supply voltage change, input voltage levels, and age. Some systematic errors may appear random with respect to one another inasmuch as their causes are independent. For instance, a signal-conditioning amplifier might have an input bias current that changes as follows:

$$-0.01 \text{ percent for } \Delta \text{ temperature} = +30° \text{ C, and}$$
$$+0.02 \text{ percent for } \Delta \text{ supply volts} = +5 \text{ percent.}$$

These effects are not only opposite but have been caused by effects that are totally unrelated.

Statistical or Random Errors

Statistical or random errors are usually a product of an arbitrary selection of a manufacturer's components and parts. For example, out of a thousand operational amplifiers of a given type, some will have positive offset voltages while some will have minus. In the multiplexer switch example we considered earlier, some will have R_{on} values that are greater than the mean, and some will have values that are less. Some R_{on} values will be within $\pm 1\ \Omega$ from the mean, and some will be greater. If not tampered with, the distribution of parameters will follow the normal curve. Like systematic errors, random error effects can be caused by changes in operating conditions. The difference is that the result is positive for some components and negative for others, rather than always in one direction.

The practical way to combine all three types of errors is as follows:

1. Add all average value or mean value errors algebraically.
2. Add all systematic error values due to specific causes algebraically to form parametric error functions.
3. RSS all random errors with the parametric values calculated in entry 2 using the maximum values of entry 2 as 3σ values. Utilize the 3σ values of the random errors or the maximum deviations where the 3σ values are not available.
4. Finally, combine the error sources in the following manner:

$$\text{Error}_{\text{total}} = \sum_{n=1}^{n} \epsilon_{an} + \left[\sum_{m=1}^{m} \epsilon_{rm}^2 + \sum_{s=1}^{s} \epsilon_{ps}^2 \right]^{1/2} \tag{9–4}$$

where ϵ_{an} = average error due to cause n
ϵ_{rm} = random error due to cause m
ϵ_{ps} = parametric error due to cause s

The one error source we have not dealt with as yet is quantization error. Although random in nature, when not coupled with noise, its distribution is rectangular rather than normal. If the quantization error is significant with respect to the other system errors, the distributions would have to be added point by point. Since doing so is difficult, the recommended approach is to combine all the analog errors, as shown, and to carry the quantization error as a separate factor. The final result is of the form:

$$\underbrace{y\%}_{\text{Average Error}} \pm \underbrace{x\%}_{\text{Random and Systematic Error}} \pm \underbrace{\tfrac{1}{2}\text{LSB}}_{\text{Quantization Error}} \qquad (9\text{–}5)$$

As an example of how the errors for a typical data acquisition system might accumulate, see Table 9-1. The system evaluated has an analog multiplexer, a signal conditioning amplifier, and a 12-bit A/D converter. The error would be accumulated and combined as follows:

1. Add the average errors algebraically:

$$\sum_{n=1}^{n} \epsilon_{an} = +0.01\% + 0.04\% + 0.02\% + 0.02\% = 0.09\%$$

2. Add all systematic values due to specific causes algebraically:
Due to temperature:

$$\epsilon_{P(T)} = \pm 0.02\% \pm 0.01\% \pm 0.005\% \pm 0.01\% = \pm 0.045\%$$

Due to power supply:

$$\epsilon_{P(PS)} = \pm 0.005\% \pm 0.01\% \pm 0.01\% \pm 0.01\% \pm 0.01\%$$
$$= \pm 0.045\%$$

3. RSS all random errors with the parametric systematic errors, where:

$$\sum_{m=1}^{m} \epsilon_{rm}^2 = 0.002525(\%)^2$$

Table 9-1 Error Analysis Example

Error Source	*Average Errors (%)*	*Random Errors (%)*	*Systematic Errors (%)*	
			Temperature ($\Delta = \pm 25°$ C)	*Power Supply* ($\Delta = \pm 5\%$)
1. *Analog Multiplexer:*				
R_{on}	+0.01	±0.01		
Off-channel leakage	+0.04	±0.02	±0.02	±0.005
V_{offset}	—	±0.01		
Settling time	—	±0.01		
2. *Amplifier:*				
V_{offset}	—	±0.01		
I_{bias}	+0.02	±0.01	±0.01	±0.01
Gain	—	—	±0.005	
CMR	—	±0.02		
Power supply rejection	—	±0.02		±0.01
Settle time	—	±0.01		
3. *A/D Converter (Analog):*				
V_{offset}	—	±0.01		
I_{bias}	+0.02	±0.01	±0.01	±0.01
Power supply rejection	—	±0.02		±0.01
Settle time	—	±0.01		
Aperture	—	±0.005		
4. *Quantization Error:*		±0.0125		

and:

$$\sqrt{\underbrace{(0.002525)}_{\text{Random}} + \underbrace{(0.045)^2}_{\text{Temperature}} + \underbrace{(0.045)^2}_{\text{PS}}} = \pm 0.081\%$$

4. Combine all error sources:

$$\text{Error}_{\text{total}} = + \underbrace{0.09\%}_{\substack{\text{Mean}\\ \text{or}\\ \text{Average}}} \pm \underbrace{0.081\%}_{\substack{\text{Random}\\ +\\ \text{Systematic}}} \pm \underbrace{0.0125\%}_{\pm\frac{1}{2}\text{LSB}}$$

To keep the problem within practical bounds, only a subset of potential error sources were used. A more complete list of potential error sources can be found in Table 9-2.

9-5 AUTOMATIC CALIBRATION

Earlier in the chapter, we mentioned that the average system error can often be adjusted out by an initial calibration of the equipment at room temperature. Techniques also exist for the compensation of systematic errors. One of the most notable of these techniques is *compumetric calibration*. This concept capitalizes on the fact that either the data acquisition system contains a computer or the measurements it makes will eventually be processed by one.

Two circuits that allow compumetric calibration are seen in Figure 9-8, with the simpler one in Figure 9-8a. The circuit requires that two extra analog channels be provided in the multiplexer. One provides a zero input, while the other provides a full-scale signal level. The full-scale value is supplied by a precision voltage source, while the zero input is provided by shorting an input pair.

The two channels are sampled either every multiplexer cycle, every few cycles, once every several minutes, or as often as is needed to maintain the system calibration. When the zero channel is sampled, an offset error correction is generated. When the full-scale channel is sampled, a gain correction factor is developed. To apply the corrections, a linear equation of the form $a(x + y)$ is used, where a is the gain correction, x is the measured value, and ay is the offset correction.

A second technique for compumetric calibration can be seen in Figure 9-8b. Here each channel contains an amplifier, preceded by a double-pole, double-throw relay. The common relay is used to first

Table 9-2 Potential Error Sources

Error Term	*Average Errors*	*Random Errors*	*Systematic Errors*			
			Temperature	*PS Volts*	*Input Volts (Differential)*	*Frequency*
1. *Source or Transducer*:						
a. Source loading	✓	✓				
b. Deviation of R_S	✓	✓	✓			
c. Source capacitance		✓				
d. ac response		✓				✓
e. Input volts—cumulative						
2. *Mux or Scanner*:						
a. R_{on}	✓	✓	✓	✓	✓	
b. V_{offset}	✓	✓	✓	✓		
c. $I_{leakage}$	✓	✓	✓			
d. ac response		✓			✓	✓
e. Settle time		✓			✓	
f. Crosstalk		✓			✓	✓
g. Sampling accuracy	✓					✓

3. *Signal Conditioning*:						
a. R_{in}	✓	✓				
b. V_{offset}	✓	✓				
c. I_{bias}	✓	✓				
d. R_{out}	✓	✓				
e. Linearity	✓		✓	✓	✓	
f. Gain	✓		✓	✓	✓	
g. Settle time		✓				
h. ac response		✓				✓
i. Z_{in}		✓				
j. Z_{out}		✓				
k. Aperture (S/H)	✓					
4. *A/D Converter*:						
a. Z_{in}		✓				
b. I_{bias}	✓		✓	✓		
c. V_{offset}	✓		✓	✓		
d. Linearity			✓	✓	✓	
e. Quantization		✓				
f. ac response		✓				✓
g. Settle time		✓				
h. Sample accuracy	✓					✓
i. Aperture	✓					✓

short all inputs and then to connect all inputs to the full-scale reference voltage. Although the same linear expression is used for calibration, this time each individual channel may be independently adjusted.

Sometimes, the data acquisition system has its own internal correction loop, as seen in Figure 9–9. Periodically, switch S1 is connected to the fixed resistor, nominal R_S. After the switch has settled, the A/D converter measures the combined signal conditioner and A/D converter offset voltage, and the result is set into the holding register. The number of bits (m) set into the holding register are just enough to handle the largest expected error. The error is D/A converted and applied to the signal conditioner in a sense that provides compensation. S1 is then switched back to the input measuring mode.

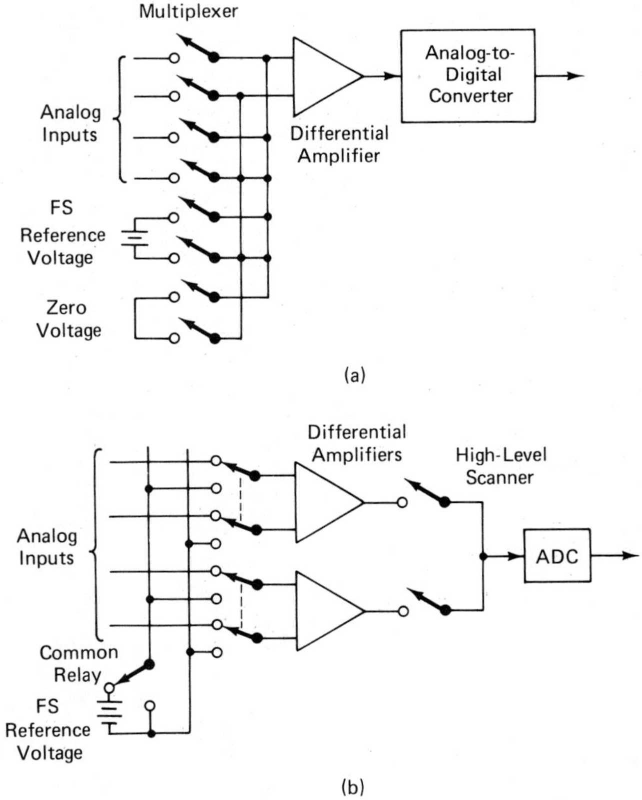

Figure 9–8. Circuits to Aid Compumetric Calibration

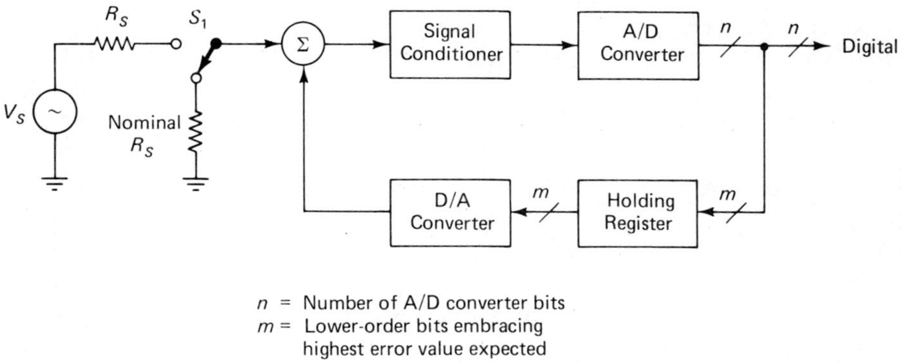

Figure 9-9. Correction Loop

PROBLEM

9–1. A data acquisition system contains an analog multiplexer, a signal-conditioning amplifier, and a 10-bit A/D converter. The multiplexer channel is connected to transducers that all have the same characteristics. The error effects for each of the components of the system are as follows:

Error Source	*Average Errors (%)*	*Random Errors (%)*	*Systematic Errors (%)*	
			Temperature ($\Delta = \pm 25°$ C)	*PS Variable* ($\Delta = \pm 5\%$)
1. *Source*:				
R_S	+0.02	±0.01		
2. *Analog Max*:				
V_{offset}	+0.01	±0.01		
$I_{leakage}$	−0.02	±0.01	±0.03	±0.01
R_{on}	+0.01	±0.01		
T_{settle}	—	±0.01		
3. *Signal Conditioner*:				
V_{offset}	+0.01	±0.01		
I_{bias}	+0.03	±0.02	±0.01	±0.01
Gain	—	±0.01	±0.01	
CMR	+0.01	±0.01		
PS rejection	—	±0.03		±0.02
T_{settle}	—	±0.01		

4. *A/D Converter*:				
V_{offset}	−0.01	±0.01		
I_{bias}	+0.01	±0.01	±0.02	±0.01
PS rejection	—	±0.01		±0.01
T_{settle}	—	±0.02		
Aperture	+0.005	±0.01		
5. *Quantization*:	—	±0.1%		

Compute the system accuracy and express it in the form of expression (9-5).

REFERENCES

1. D. F. Hoeschele, *Analog-to-Digital/Digital-to-Analog Conversion Techniques*. New York: John Wiley & Sons, Inc., 1968.
2. H. Schmid, *Electronic Analog/Digital Conversion*. New York: Van Nostrand Reinhold Company, 1970.
3. W. Myers, "Uncover Data Acquisition Errors," *Electronic Design* (May 10, 1975).
4. B. Gordon, "What's Wrong with A/D Converter Specs?," *EEE Magazine* (February 1969).

10 Timing and Control

10-0 INTRODUCTION

Up to this point in the text, we have either concerned ourselves with the analog portions of the data acquisition system or investigated the processes involved with the translation between the analog and the digital domains. In this chapter, let's consider the digital side of the system.

Whereas we have previously been concerned with the many factors and problems affecting the system's measurement accuracy, we now investigate such things as timing, formatting, controlling, sequencing, and input/output (I/O) interfacing. These parameters are familiar ones in the realm of digital logic. This chapter does not provide a course in logic design, since many excellent texts cover the topic (see References). It does, however, provide an insight into the ways in which logic circuit elements are utilized to implement timing and control for the data acquisition system.

Data acquisition systems can be found in many different forms, varieties, and shapes, depending on the application. For the purposes of this discussion, we will divide them into two broad categories—namely, those that are peripheral to a digital computer, and those that are stand-alone (without a computer). Top-level block diagrams for the two basic categories can be seen in Figure 10-1. With the stand-alone

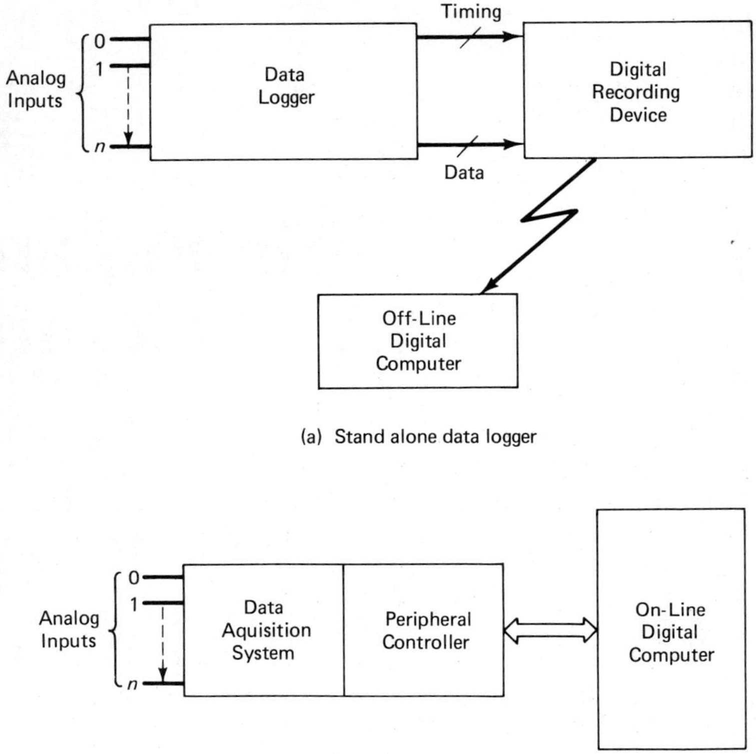

(b) As a peripheral to a digital computer

Figure 10-1. The Basic Data Acquisition System

system of Figure 10-1a, the converted data is stored on a recording medium for later computer data reduction. In this case, the data acquisition system, or data logger, contains its own master clock and provides timing to the recording device. The system shown in Figure 10-1b usually derives its timing from the on-line computer clock, and it operates on commands from the computer. To do so, a special I/O unit called a peripheral controller is needed to properly interface the data acquisition front end to the computer.

In the remainder of this chapter, we will investigate these two system types from the point of view of timing and control.

10-1 THE STAND-ALONE DATA LOGGER

The stand-alone data logger is a totally self-contained unit that can acquire data, sample and quantize it, place it into an appropriate digital format, and provide the proper interface for the selected storage devices. As such, it must contain its own timing and control system, which can be seen in Figure 10–2. Although the diagram looks much like that found for the typical data logger in Chapter 1, a number of additional lines for timing and information have been included. To establish a baseline for discussion, a number of basic assumptions about the data logger's characteristics have been made:

1. *Number of analog channels*: 100 first-channel/last-channel selection and display.
2. *Signal-conditioning amplifier gain steps*: 4 (automatic, 100-mV, 1 V and 10 V full scale).
3. *A/D converter resolution*: 4 BCD digits and sign (with display).
4. *Manual data entry*: 3 BCD digits.
5. *Timing-of-day clock*: Hours, minutes, seconds in BCD (with display).
6. *Data code*: ASCII* numeric, with time, manual data, and channel number as options.
7. *Operating modes*: 5-continuous (CO), single-scan (SS), single-step (ST), single-channel (SC), and manual.

Each of these items is explained more fully as the investigation proceeds. To effect a cohesive flow, the discussion will start at the multiplexer and move through the system to the output buffer.

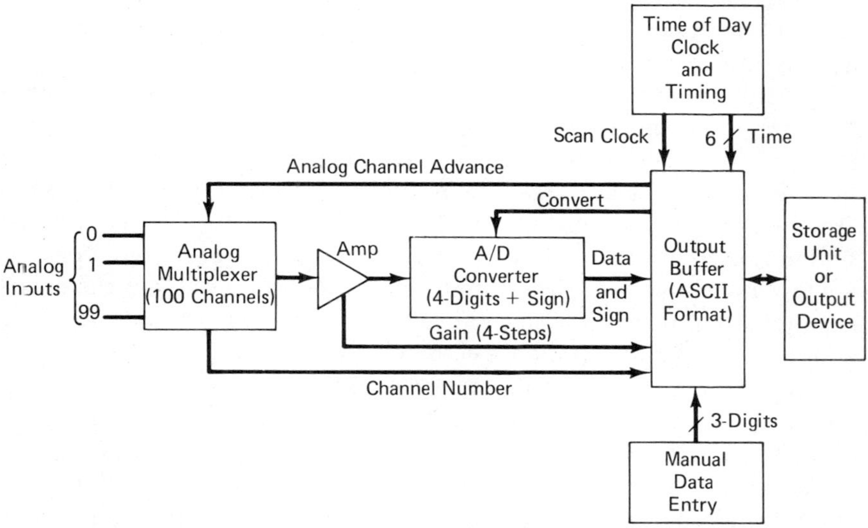

Figure 10–2. Example Data Logger Block Diagram

*American Standard Code for Information Interchange.

10-1.1 The Input Section

The input section of the data logger can be seen in Figure 10-3. It consists of the analog multiplexer and the signal conditioning amplifier. Let's consider each of these in turn. As the diagram indicates, the analog multiplexer is more than just a simple switch array. It also contains the drivers to turn the switches on and off, as well as a control section to allow the sequencing of the switches in the manner desired. For our example, the switches will be activated in numerical sequence from a manually selected first channel to a manually selected last channel.

A more detailed look at the multiplexer control section can be seen in Figure 10-4. At the heart of the section are the control logic and counters, since they control the state or position of the multiplexer. Bounded by the first-channel and last-channel selection, the counters sequence from one channel to the next, as channel advance signals are received from the DAS output buffer section (Figure 10-2).

Figure 10-5 is a functional logic diagram of the first-channel/last-channel select logic. Under normal conditions, the channel select register/counter is in the counter mode, and it is advanced by the analog channel advance via the AND gate, AND(1). Whenever the counter output matches the channel number contained in the last-channel select, a positive compare is detected by the exclusive-OR last-channel detector. This action shuts off AND(1) and activates AND(2), placing the register/counter in the register mode. The next analog channel advance input does not sequence the counter but rather transfers the first-channel select output into the register. At this point, the last-channel detector no longer detects a compare, thereby switching off AND(2) and reactivating AND(1).

After the system measures the first-channel input, the next analog channel advance input once again sequences the counter. When the last channel is reached, once again the process repeats. As seen in both Figures 10-4 and 10-5, a numeric channel number display is attached to the counter output. This display is helpful whenever the system is used in the manual mode to allow the user to see which channel is being converted. Referring to Figure 10-4, we see that the channel number information is also available to the output buffer, along with the last-channel compare signal.

One other section of Figure 10-4, the 2-digit to 100 decode, has not been discussed. The output of the counter is in binary-coded decimal (BCD) form, which is not directly usable by the analog switches. Logic must be provided to convert the counter codes into 100 distinct and separate outputs in order to activate the individual switches.

Figure 10-6 shows one of the ways to implement the decode logic. The multiplexer select counter consists of two sections: one for the

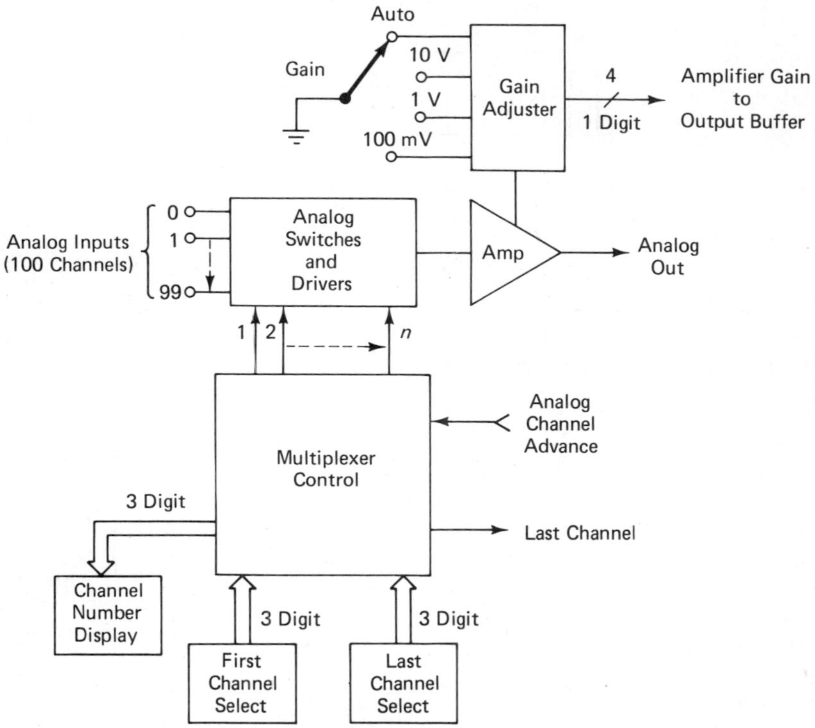

Figure 10–3. Block Diagram of Analog Multiplexer and Signal Conditioning Amplifier

one's decade, and the other for the ten's decade. The other key element is a 4- to 10-line decoder. This decoder converts the BCD numbers from 0 through 9 to 10 discrete outputs on 10 separate lines. The one's counter outputs are connected to the four line inputs of ten 4- to 10-line decoders, while the ten's counter has its output converted into 10 individual outputs by a separate 4- to 10-line decoder. The outputs from this decoder generate the signals enable 1 through enable 10.

The system operates as follows. When the counter is reset (all outputs are zero's, indicating channel "0"), the enable-1 line is activated, thus enabling decoder 1. As the one's counter sequences from 0 to 9, the decoder's outputs cycle from 0 to 9. In this fashion, the analog multiplexer switches 0 through 9 are activated, in order. As the one's counter carries to the ten's counter on the ten count, it resets itself back to 0. Now enable 1 is deactivated, and enable 2 is activated. This time, when the one's counter counts, analog channels 10 through 19 are sampled. This process continues until all hundred channels (0–99) have been exercised. Should the first-channel setting be other than 0 and the last-channel setting other than 99, the first channel/last-channel

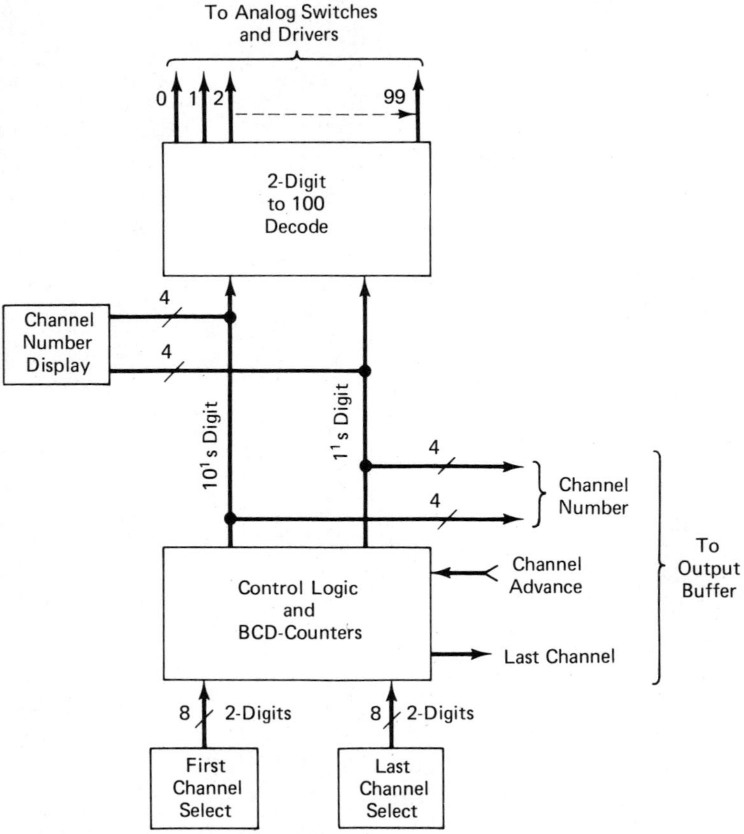

Figure 10-4. Multiplexer Control Section

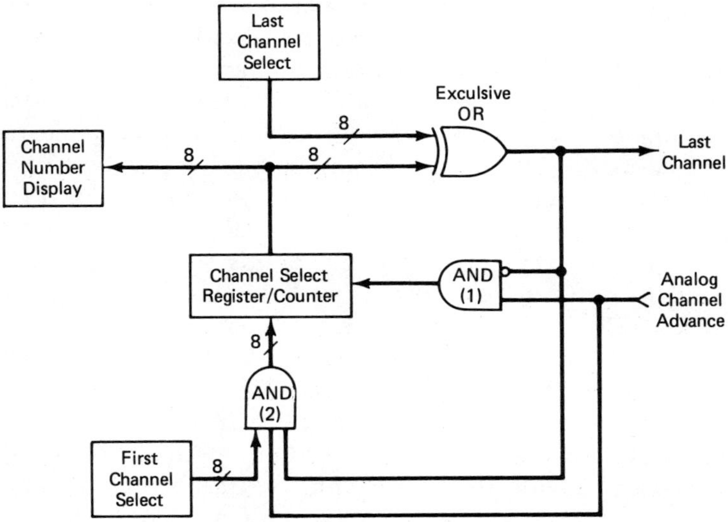

Figure 10-5. First-Channel/Last-Channel Select Logic

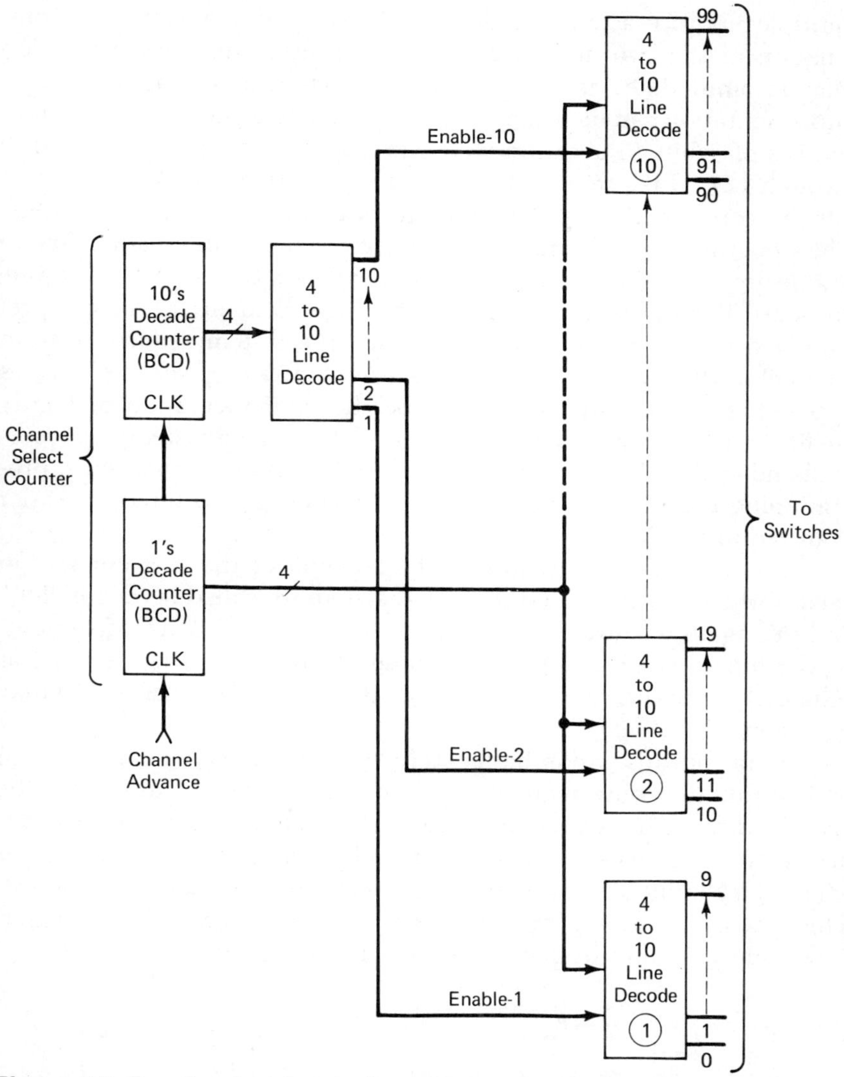

Figure 10-6. Control Logic for 100-Channel Multiplexer

select logic causes the counters to cycle between the selected channels.

So far, we have considered a multiplexer front end based on the binary-coded decimal number system that our example data logger called for. Digressing for a moment, we could just as well have used the standard binary system, which the digital computer more readily understands. Such a system can be implemented as easily as the BCD system just discussed. In fact, 3-bit binary to 8-line decoders and 4-bit binary to 16-line decoders, as seen in Figure 10–7, are readily available. Such devices allow simple conversion from 3- and 4-bit binary to 8 and 16 discrete lines. With a binary counter and one of these devices, multiplexers with a few channels can be easily implemented. A combination of the two devices can yield a multiplexer implementation that accommodates as many as 128 discrete analog channels. This implementation can be seen in Figure 10–8. The channel select counter consists of 7 bits to accommodate the required 128 possible states (0 through 127). Like the BCD system of Figure 10–6, the lower-order 4 bits are connected to the 4-line inputs of eight 4- to 16-line decoders. The uppermost significant three bits are converted into eight discrete enable lines (enable 1 through enable 8) by a separate 3- to 8-line decoder. The operation of this control logic is identical to that previously discussed for Figure 10–6, since the system sequences from channel 1 through channel 128. If the computer prefers binary as opposed to BCD, why would you use the implementation of Figure 10–6? As it happens, a large portion of data logging systems are operated and utilized by noncomputer-oriented people. For their sake, programming the computer to convert a BCD output to binary is easier than requiring them to convert binary to decimal.

Another point of interest is the labeling of the first channel in each of the two systems. For the BCD system we called it "channel 0," and this again is convention for the human-operated BCD instrument. In the binary situation, we have chosen to label the first channel as "channel 1." In this case, there is no convention, and either option may be exercised.

So far our discussion has centered on systems whose channels can be commutated only sequentially. The only way a channel may be examined more than once per cycle with this type of system is to connect it to more than one DAS input. There are, however, systems in which the channels are not necessarily selected in numerical sequence. This type of situation occurs when the analog sources to be measured have widely varying frequency content from channel to channel.

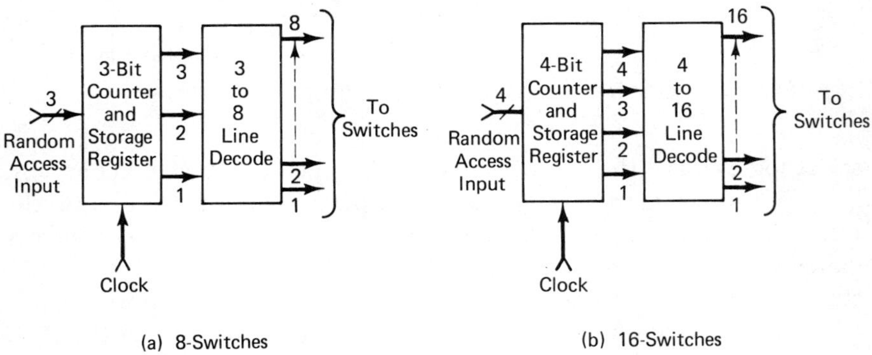

Figure 10–7. Control Logic for Small Number of Switches

A scan sequence for multilevel commutation can be seen in Figure 10-9. Twelve channels are scanned in accordance with a sequence that consists of sixteen 1-sec time segments or cycles. The channels that are sampled once per second are assumed to be the baseline (called "commutation"). The channel sample rates are labeled as follows:

Channels Commutated	*Sample Rate*	*Label*
1	3 × per sec	Super commutation
3, 4, 5, 6	1 × per sec	Commutation
2, 7, 8	1 × per 4 sec	Subcommutation
9, 10, 11, 12	1 × per 16 sec	Sub-subcommutation

Solving the supercommutation problem is fairly straightforward. Multiple outputs of the multiplexer selection decoder are ORed together at the supercommutated channel driver input (channel 1 in this case). The subcommutated and sub-subcommutated channels are not so simple, because additional counters must be used to keep track of which cycle the sequence is in and where the commutator is within the cycle.

One area of discussion remains within our example data logger input section. Referring back to Figure 10-3, we find that we have not yet considered the signal-conditioning amplifier. In our example, we have called for four gain-setting positions: automatic, 10 V full scale, 1 V full scale, and 100 V full scale. Each of these positions is manually selectable on the front of our system. Should all 100 inputs be comfortably within any one of the last three scales, the switch can be set to that position. The one-digit (four-line) output to the output buffer always indicates that setting. Yet if the channels could best be measured by utilizing a number of ranges, the system would be set to automatic. In the automatic position, the range automatically changes to the correct one when the input magnitude is detected. The one-digit gain output is again available to the output buffer for recording. DAS operators must recognize that gain ranging requires additional time as the system moves from channel to channel, and they must account for it when selecting scan rates.

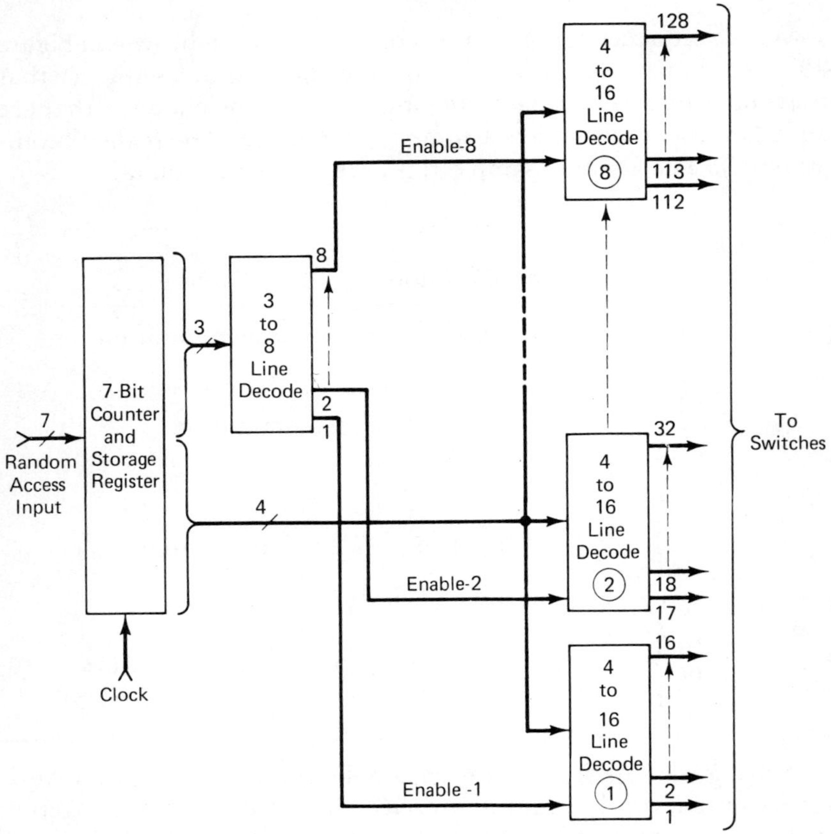

Figure 10–8. Control Logic for 128-Channel Multiplexer

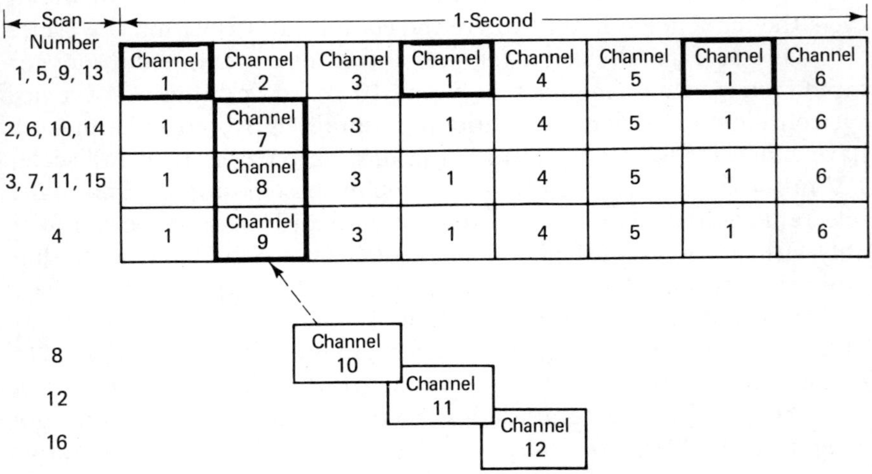

Figure 10–9. Pattern for a 16-by-1-Second Scan Sequence

10-1.2 The Conversion Section

The data acquisition system conversion section block diagram can be seen in Figure 10-10. Its timing and control are fairly simple. At an appropriate time when the newly selected analog channel has settled to within the required accuracy, the sequencer position of the output buffer (Section 10-14) issues an A/D convert command. The A/D converter performs the required quantization and presents the answer (4 digits of data plus a sign digit) to the output buffer's data selector. Data and sign are also presented for numeric display. Along with the channel number display, this allows for the manual step and calibration of each channel. As operators step the unit from channel to channel, they can directly read the results of the measurements.

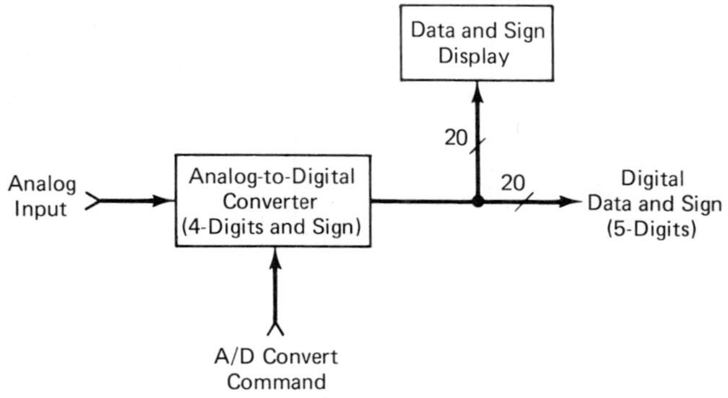

Figure 10-10. Analog-to-Digital Conversion Section

10-1.3 Timing and Clock Generation

At the heart of the data logger is the master system clock, as well as the circuitry that determines the sample and scan rates. The good DAS designer ensures that everything that happens within the system is synchronized to the master clock, whenever practical.

Very often, the data logger also offers a time-of-day clock as a feature, which, as it happens, is part of our example system. This feature allows the user to insert the time of day that the data acquisition run was made, along with the data recorded from the run. The time-of-day clock for our system can be found in Figure 10-11.

The clock is a series of counters running off the master 100-kHz system clock. The first five decades (dividing by 10) count the clock down by 10^5 to an output that occurs once per second. The next decade counter counts seconds, while the one that follows is short-

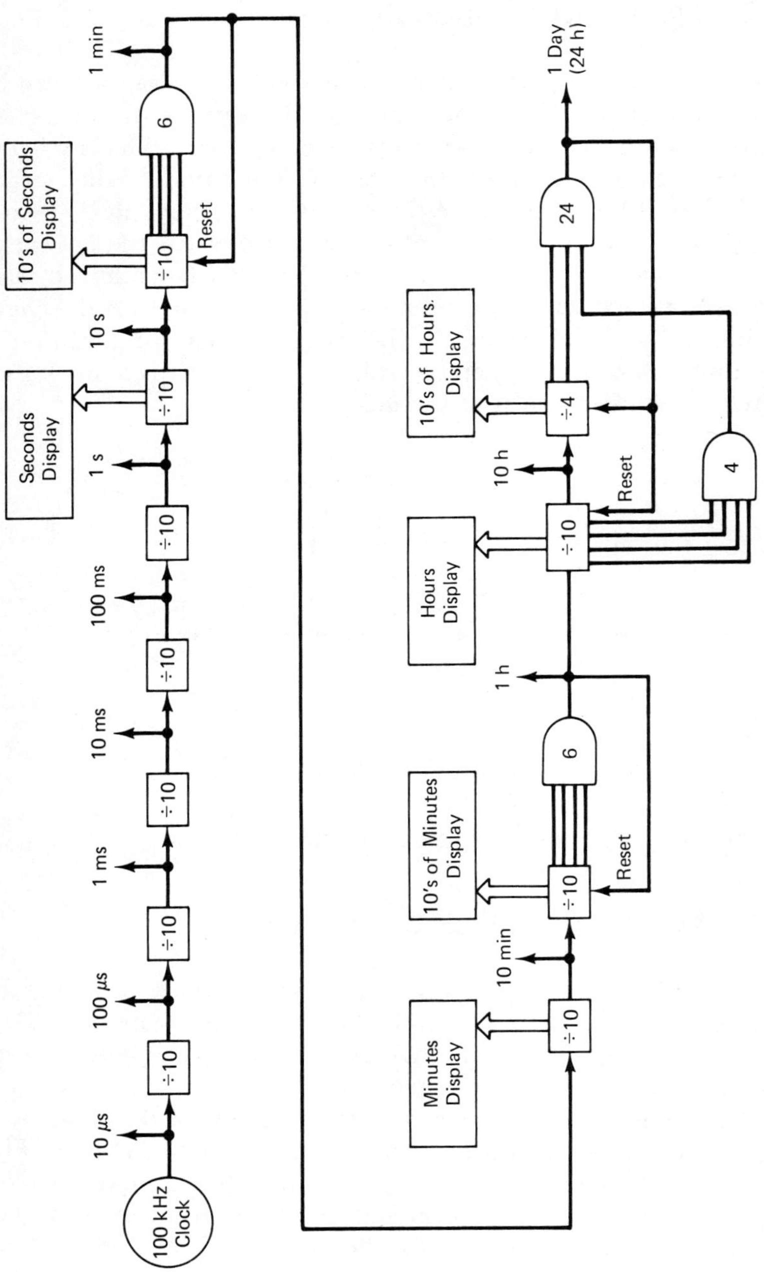

Figure 10–11. Time-of-Day Clock

counted (divided by 6) to provide the ten's of seconds or one minute out. The same situation exists for the next two decades, but their outputs are minutes and ten's of minutes. Finally the last two counters are interconnected to register hours and ten's of hours, up to 24. The outputs of the last six counter stages are displayed for operator use and for setting purposes. The same six digits of time information are also available to the data logger output for recording on demand.

Although our example data logger does not require it, systems commonly provide the day of the year as well. In this case, the output of the hours counter set (which reflects one day) is provided to an additional counter set, such as that shown in Figure 10-12. Here the counters and their accompanying logic are interconnected to cycle from 0 to 365 days and to provide outputs for days, tens of days, and hundreds of days. Once again the outputs of the counters are made available for both display purposes and for recording with the data. With this system, operators can have a permanent record of the day and time when their measurements were accomplished. A standard accuracy specification for our clock would be "±0.01% @ 25° C ± 0.005% per °C." If the system did not encounter any prolonged and severe temperature excursions, we would expect less than 10 sec of error a day and less than an hour per year, which is quite sufficient for most DAS situations. So much for establishing the time that the data logger performs its measurements.

Internal system timing is also a problem to be solved. A circuit to satisfy these requirements can be seen in Figure 10-13. The outputs from the time-of-day clock are combined by the scan rate selection logic to form the various scan clocks for the output buffer. A *scan* is

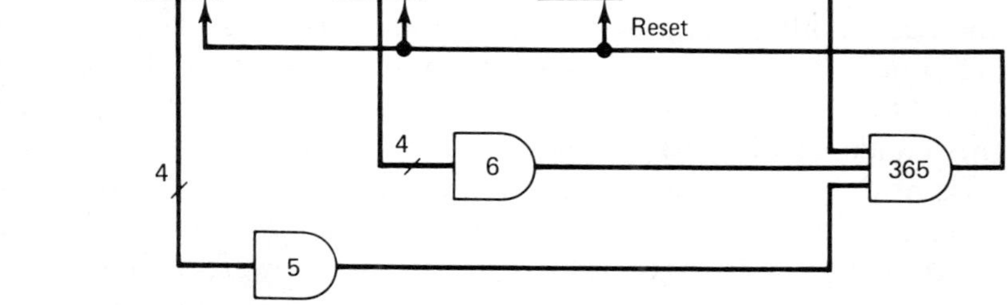

Figure 10-12. Day-of-Year Clock

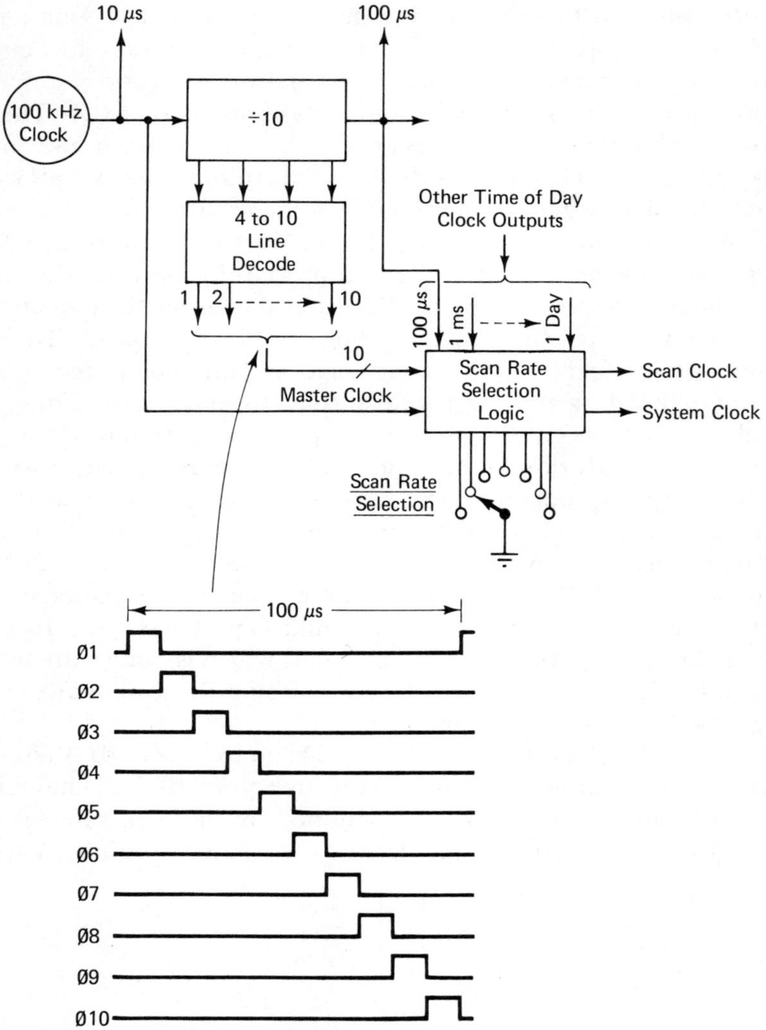

Figure 10-13. Internal Housekeeping Clock Generation

defined as a multiplexing cycle that goes from the first channel to the last channel. Scan rates can range from a scan of the set of multiplexer channels once per day to measurements of a new channel every few microseconds. An important feature of the clock generation shown is the use of a 4- to 10-line decoder, attached to the first decade. This decoder provides ten phases of 100-kHz clocks, as shown in the diagram to the left of the figure. Multiple clock phases greatly reduce the problems involved in complex timing generation. An added feature is that the total system can be synchronized.

10-1.4 The Output Buffer

The output buffer is the data collection point for the data logger. As you can see in Figure 10-14, the outputs from all functions, which have data available for recording, have inputs to the output buffer. The information is placed in proper format and prepared for storage on the output device medium. The buffer provides the proper data code (ASCII, in this case), as well as the drive and control signals required to interface to operate the device.

A more detailed view of the output buffer can be seen in Figure 10-15. It contains four major functions:

1. the sequencer,
2. the record-length counter,
3. the data selector, and
4. buffer drivers and receivers.

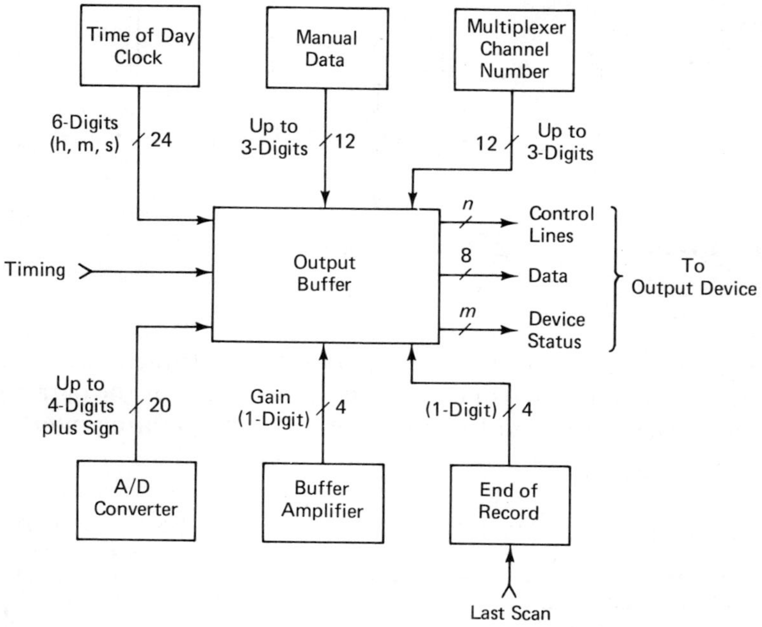

Figure 10-14. Output Buffer Interfaces for Typical Data Logger

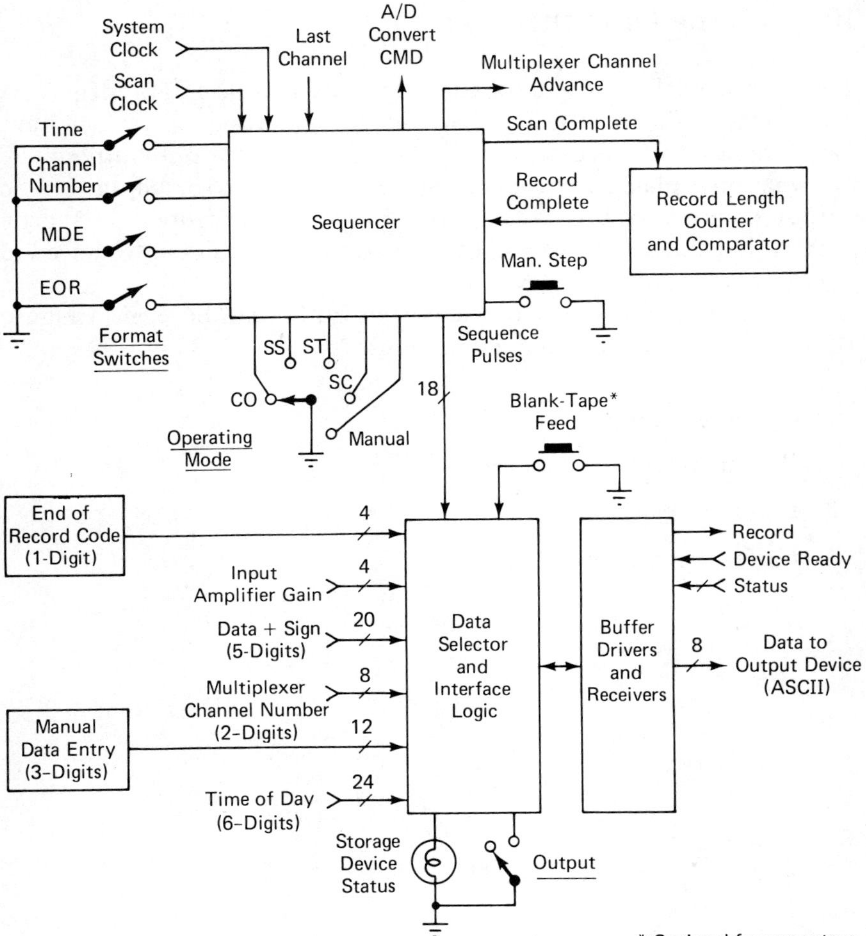

* Optional for paper tape.

Figure 10-15. Output Buffer Functional Diagram

The sequencer is the most complicated of the four functions. Here the signals are generated to sequence the data generated by the data logger into the output device, utilizing the appropriate operating mode. The recording sequence for our example data logger is:

HEADER	FIRST CHANNEL	SECOND CHANNEL	. . .	LAST CHANNEL	END OF RECORD
TT TT TT MMM	*CC ± DDDDG*	*CC ± DDDDG*		*CC ± DDDDG*	*E*

where T = time (hours, minutes, and seconds)
M = manual data
C = channel number
$\pm$ = data sign
D = data
G = amplifier gain
E = end of record

As Figure 10–15 indicates, the time, manual data entry, channel number, and end of record are all selectable and therefore optional. The sequencer must also contend with five operating modes, as follows:

1. *Continuous (CO)*: The multiplexer continuously scans from the first to the last channel at the maximum system throughput rate. Whenever the system is changed from this mode to a nonoperate mode (such as manual), it completes the present scan before stopping. If the time and manual data switches are set to on, time and manual data are recorded at the beginning of each scan.
2. *Single-scan (SS)*: A single scan (from first to last channel) is accomplished each time a scan clock is received from the timing section. Whenever the operating mode is changed, the system completes the present scan. Time and manual data are optionally recorded at the beginning of each scan.
3. *Single-step (ST)*: The analog multiplexer steps from one channel to the next each time a scan command is received. Time and manual data are optionally recorded prior to the data from the first channel.
4. *Single-channel (SC)*: The multiplexer remains on the channel selected by the first channel selector and continues to record the data from the single channel.
5. *Manual (Man)*: The multiplexer is advanced from first channel to last channel, one channel at a time, each time the manual-step button is depressed. The channel number and results of the A/D conversion are available on numeric displays.

Utilizing combinations of the operating modes and the various scan clock rates from the timing section, you may generate almost any scan rate from once a day to the maximum supported by the output device.

At appropriate points in the sequence, the sequencer generates A/D conversion and analog multiplexer channel-advance commands.

As each scan is completed, it also provides a scan-complete signal to increment the record-length counter.

There are several ways to implement the sequencer function. No matter which is chosen, the outputs must look like that shown in Figure 10-16. You can readily see that the sequencer pulses directly track with the output format previously discussed. From the diagram, you also see that the first A/D conversion in each scan is commanded immediately after the last manual data entry digit. Conversions for successive channels are commanded after the recording of the gain digit from the previous channel. The analog multiplexer is advanced after the first data digit of each channel has been recorded.

Now consider Figure 10-17. Here the sequencer is implemented with shift registers. After the occurrence of the scan-start input, which places a 1 in the first stage of the shift register, the 1 is shifted through successive stages by the shift clock. Both scan-start and the shift clock are generated by combining the operating mode setting with the system clock and scan clock from the timing section. These signals determine how often a scan should occur and at what rate the sequencer should enable the various data digits at the data logger output. Outputs from the various shift register stages form the selection pulses of Figure 10-16. Time, manual data entry, and channel number—optional in the output format—are switch selectable, as shown. In keeping with Figure 10-16, we see that the A/D conversion commands occur at the end of the last manual data entry pulse period and at the end of the amplifier gain period, when it is not the last channel and the system is not in the single-step mode. Channel advance occurs, as advertised, at the end of the first data digit time period. Whenever the last channel of a scan has occurred, a scan-complete signal is generated to increment the record-length counter. When the record is complete, the record-length comparator (see Figure 10-15) generates a record-complete signal and the end-of-record code (EOR) is stored on the output device media.

An alternative to the shift register sequencer can be seen in Figure 10-18. Counters with decode logic are used to generate the sequencer pulses. Although the diagram shows only the first two sections (time and manual data), the entire sequencer can be implemented in this manner. While this technique requires slightly more hardware than the shift register implementation, it is more foolproof. Care must be exercised when utilizing the shift register concept so that more than a single 1 does not appear in the register at any time. Should this occur, more than one data digit will be applied to the output device at the same time, and the output will be scrambled. Such a situation cannot occur as easily with the counter/decoder concept.

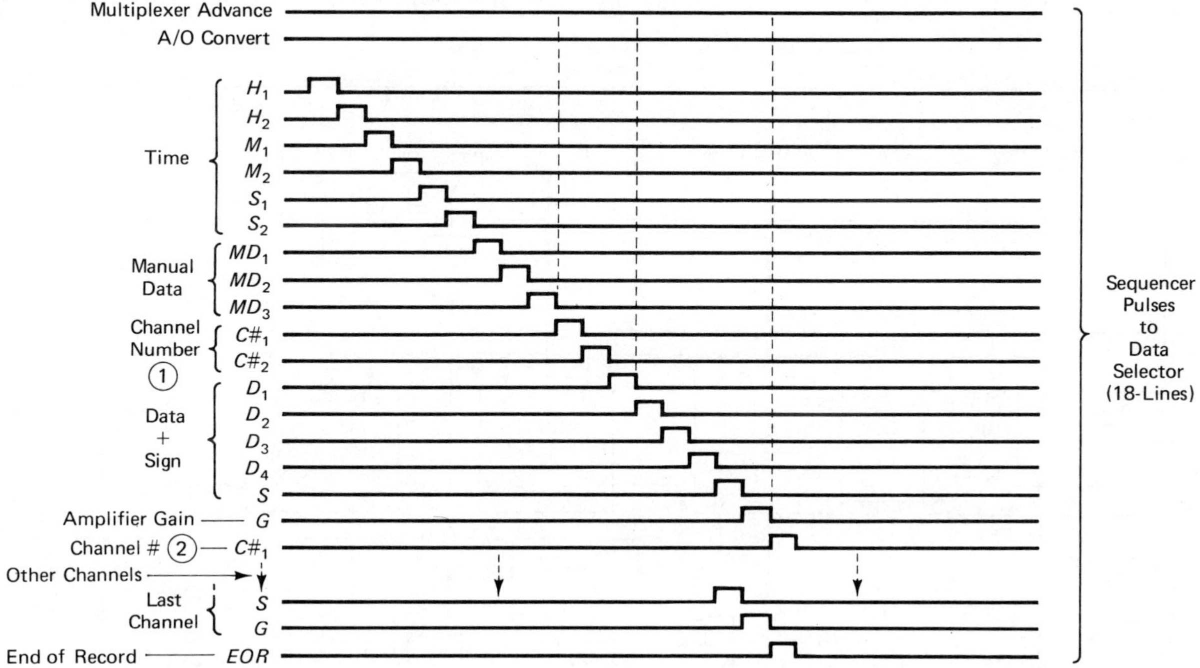

Figure 10-16. Sequencer Output Timing

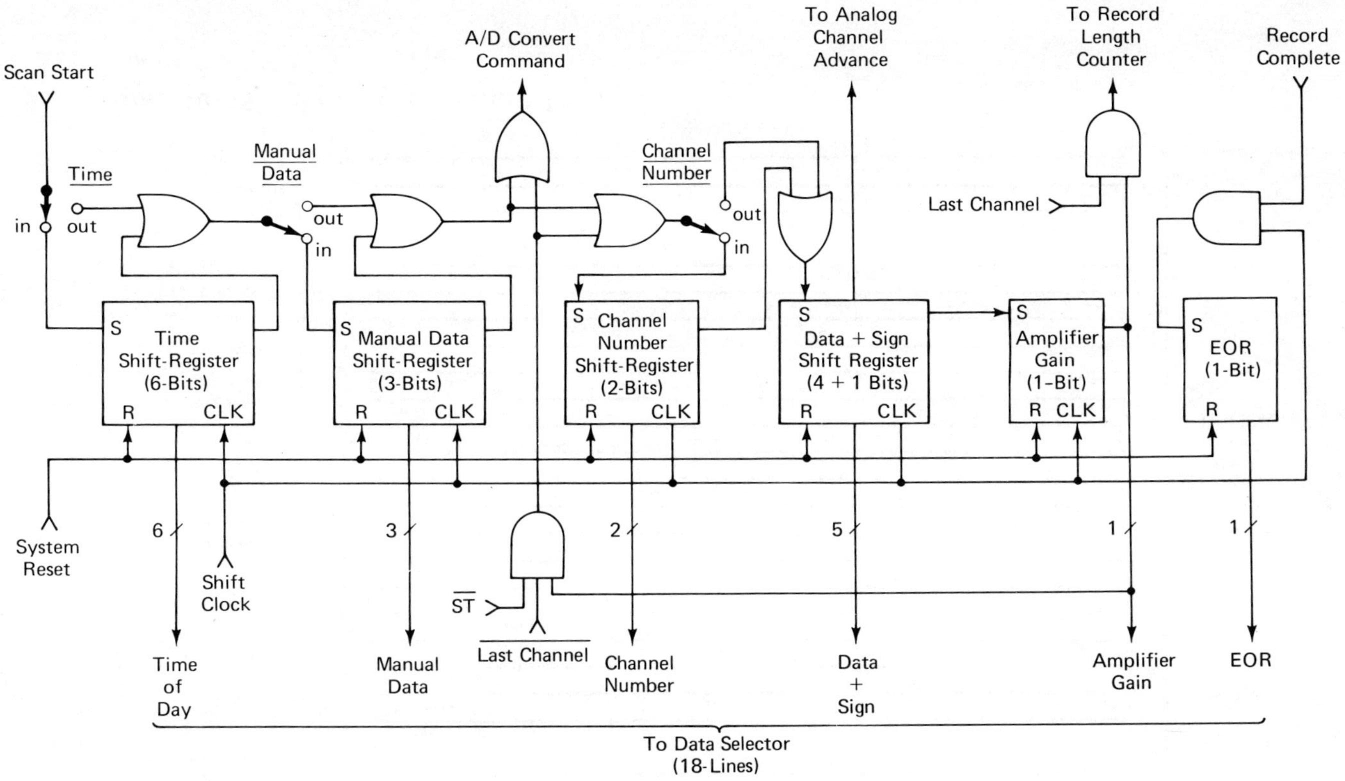

Figure 10-17. Sequencer Implemented with Shift Registers

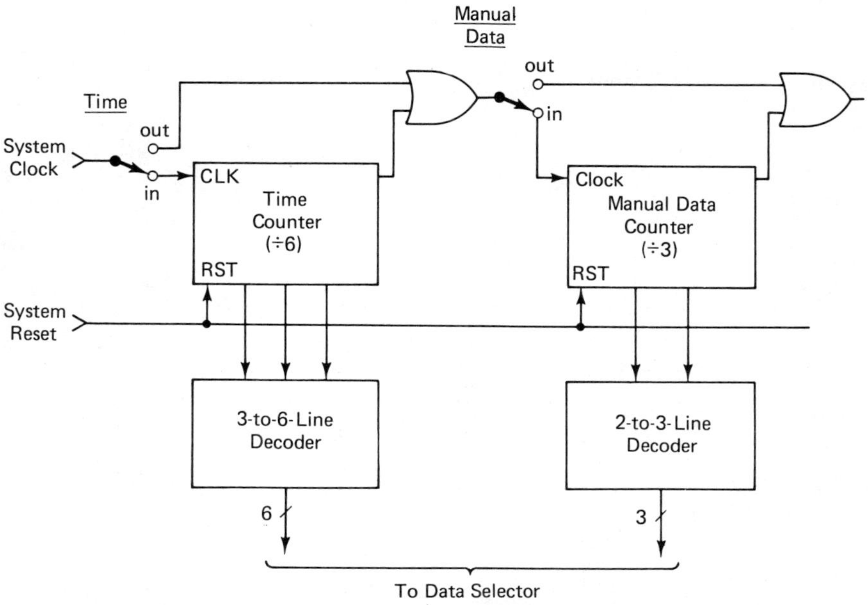

Figure 10–18. Sequencer with Counters and Logic Decode

In the late seventies, a device called a read-only memory (ROM) became popular with digital logic designers and as such also invaded the data acquisition field. The sequencer can also be implemented with this device, as shown in Figure 10–19. With this implementation, the sequencer outputs are generated as the memory is addressed or stepped through its various storage locations. The contents of the ROM, which are required for our example data logger, can be seen in Figure 10–20. Each ROM location contains the next address to be

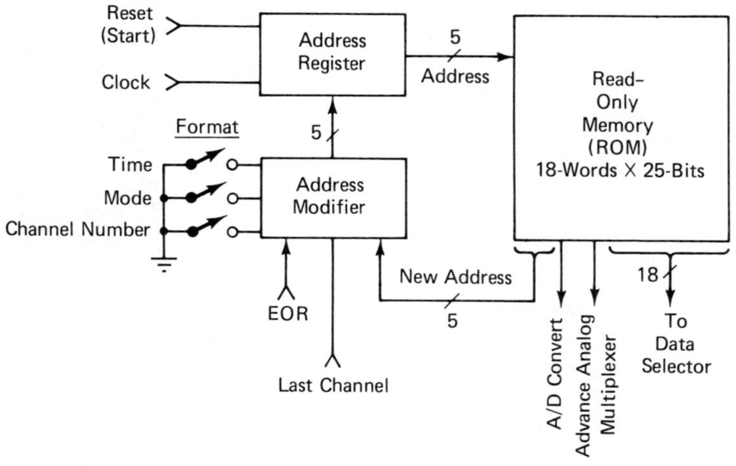

Figure 10–19. Sequencer with Read-Only Memory

ROM Location	(5-Bits) Next Address	Convert	Advanced Analog Multiplexer	Selector Output (18-Bits)
0	0 0 0 0 1	0	0	1 0 0 0 -- Enable Hours # 1 -----→ 0 0 0 0
1	0 0 0 1 0	0	0	0 1 0 0 -- Enable Hours # 2 -----→ 0 0 0 0
2	0 0 0 1 1	0	0	0 0 1 0 -- Enable Minutes # 1 ----→ 0 0 0 0
3	0 0 1 0 0	0	0	0 0 0 1 -- Enable Minutes # 2 ----→ 0 0 0 0
4	0 0 1 0 1	0	0	0 0 0 0 1 -- Enable Seconds # 1 --→ 0 0 0 0
5	0 0 1 1 0	0	0	0 0 0 0 0 1 -- Enable Seconds # 2 --→ 0 0 0 0
6	0 0 1 1 1	0	0	0 0 — — — 0 1 -- Enable Manual Data # 1 --→ 0 0
7	0 1 0 0 0	0	0	0 0 — — — 0 0 1 -- Enable Manual Data # 2 → 0 0
8	0 1 0 0 1	0	0	0 0 — — — — 0 0 1 -- Enable Manual Data # 3 -→ 0
9	0 1 0 1 0	1	0	0 0 — -→ Convert — All Zeros -------→ 0 0 0 0
10	0 1 0 1 1	0	0	0 -- Enable Channel # 1 -- 0 1 0 0 0 0 0 0 0
11	0 1 1 0 0	0	0	0 -- Enable Channel # 2 -- 0 0 1 0 0 0 0 0 0
12	0 1 1 0 1	0	0	0 -- Enable Data # 1 — 0 0 1 0 0 0 0 0
13	0 1 1 1 0	0	1	0 -- Enable Data # 2 — 0 0 0 1 0 0 0 0
14	0 1 1 1 1	0	0	0 -- Enable Data # 3 — 0 0 0 0 1 0 0 0
15	1 0 0 0 0	0	0	0 -- Enable Data # 4 — 0 0 0 0 0 1 0 0
16	1 0 0 0 1	0	0	0 -- Enable Sign — — 0 0 0 0 0 0 1 0 0
17	1 0 0 1 0	0	0	0 -- Enable Amplifier Gain -- 0 0 0 0 0 0 1 0
18	1 0 0 1 1	0	0	0 -- Enable End of Recorder — EOR — 0 0 0 1

Start → ROM Location 0

If (Last Channel) · ($\overline{EOR}$) → location 0 (from location 17)

If ($\overline{\text{Last Channel}}$) → location 9 (from location 17)

Modify Address (locations 17–18)

Stop (Next Address, location 18)

← First Convert Command Record (location 0)

← 1 Through n Convert Command/Record (location 9)

← Select New Analog Channel (location 13)

← Test for Last Channel and EOR (location 17)

Figure 10-20. Read-Only Memory Contents

accessed, an A/D conversion command output, an analog multiplexer advance enable, and 18 bits of data selector outputs. As the ROM is addressed from top to bottom, only one selector output is enabled at any time or location. This effectively produces the timing pattern of Figure 10–16. If time, manual data, or channel numbers are not desired, the address modifier of Figure 10–19 jumps the address to the appropriate ROM location. The A/D convert commands and the analog multiplexer channel advance are also programmed to occur in the proper places. The first convert command occurs when the ROM has sequenced to address 9. Successive converts, in a given scan, occur after the gain digit address at location 17. When the address modifier detects address 18 (10010) and the last-channel signal is not available, the next address is modified to force the next ROM address to be 9 rather than 18. On the occurrence of the last channel indication, when end of record has not been detected, the modifier jumps the address pointer to ROM location 0, and a new scan begins. Should address 17 be reached and end of record be indicated, the address pointer increments to location 18, and the EOR digit is recorded. The next address pointed to by location 18 is invalid, causing the sequencer operation to cease. The major advantage of this sequencer implementation is that it is easily adaptable to different operational formats by a simple change of ROM.

The Record-Length Counter

The record-length counter is an important part of the typical data logger. When analyzing data with a digital computer, it is highly desirable that the computer know where, on the recording media, the end of the recorded data resides. While this information is necessary to notify the computer that it has all its input, it can also be important for computers with limited storage capability. If the total amount of data collected is large while the computer memory capacity is small, the data must be separated into records or segments that prohibit an overload. The record-length counter allows this segmentation.

The typical data logger allows manual selection of the record length, as seen in Figure 10–21. During data logger operation, the record-length counter keeps track of the number of multiplexer scans that have occurred. When the number in the counter matches that of the manual setting, an end of record (EOR) is written on the storage device. The end of record character can have different meanings to different output storage devices. For instance, with a magnetic tape unit, the EOR character is called an "interrecord gap," while for a paper tape unit, it is the ASCII end-of-file (EOF) character.

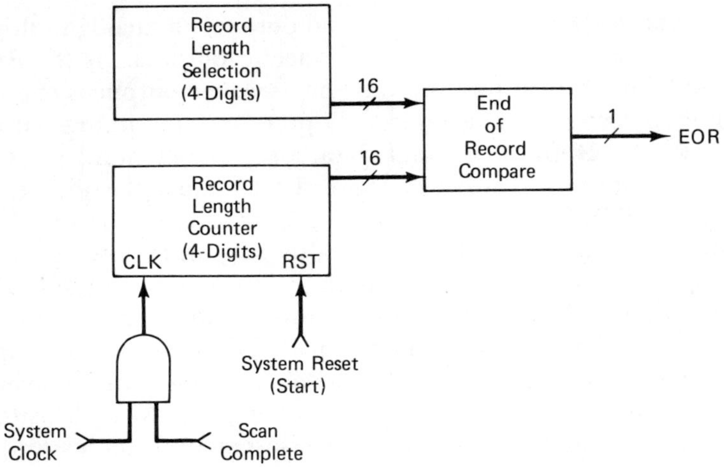

Figure 10–21. Record-Length Counter

The Data Selector, Drivers, and Receivers

If we refer back to Figure 10–15, we see that the 18 digits enabled by the sequencer are provided as inputs to the data selector. The data selector simply applies these digits to the data logger output drivers one at a time, as commanded by the sequencer.

Functional logic for a portion of the selector can be seen in Figure 10–22. When the sequencer enables the first hours line, the four bits that comprise the hours digit are multiplexed to the BCD numeric outputs. The designations of 8, 4, 2, and 1 indicate the numeric weight of the bits. As the sequencer cycles, successive groupings of AND-gates are enabled until the end-of-record (EOR) digit enable occurs. At that time, the EOR character or digit is multiplexed to the output. The hard-wired four bits shown at the top of the diagram complete the 8-bit ASCII designation for numbers (Chapter 11).

Let's look back once again at Figure 10–15 so we may consider a few remaining items. Little has been said about manual data entry, other than that it is available and can be recorded as a part of the scan header. This feature allows the insertion of identifying or qualifying data into the data record. For instance, you might wish to identify the run number or specify special test conditions that prevailed at the time of the data taking. Our example system provided for automatic insertion of manual data at the start of each scan. Other systems allow for it at the start of each record, or they permit users to manually insert as many entries as they desire before the start of the record.

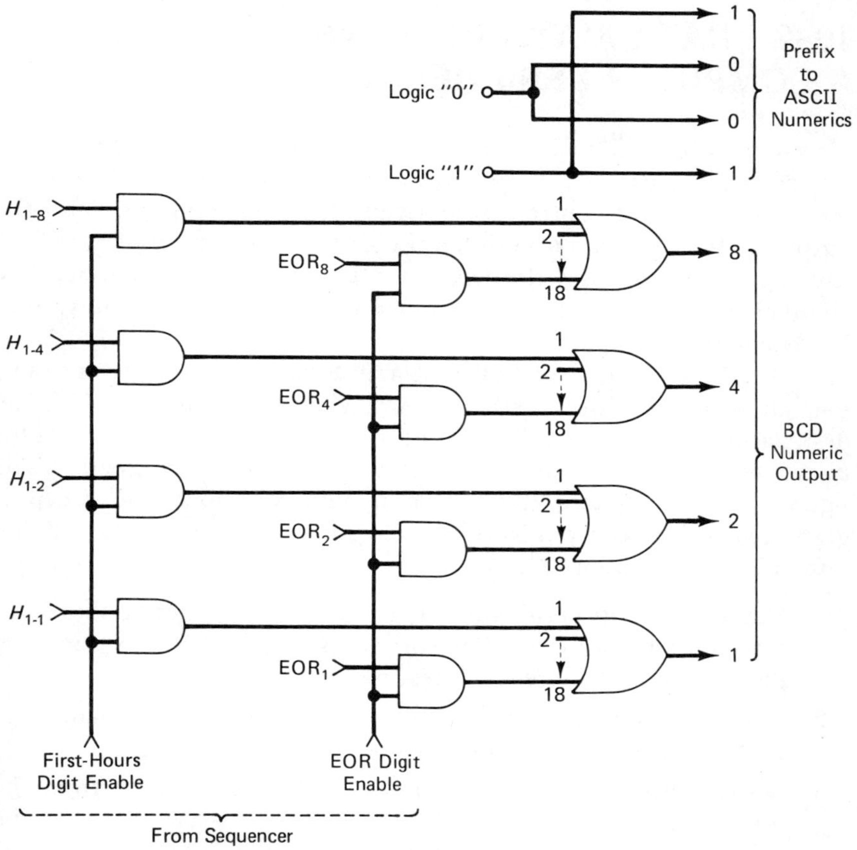

Figure 10–22. Data Selector Functional Logic

The interface logic portion of the data selector has some other features. If we are using a paper tape punch, leader may be provided by depressing the blank tape feed button. The output switch allows the output signals to the drivers to be disabled, thus allowing use of the manual-step feature of the data logger without unnecessary waste of the output medium. This capability is especially useful if the medium is punched paper tape. Often we find storage device status indicators available. For a punched paper tape system, the indicator could signal that the punch is out of tape.

There are, of course, many other implementations of data loggers, some much more sophisticated than our example, and some of greater simplicity. But regardless of the configuration, they all utilize variations of the techniques discussed in the previous sections of this chapter. So the concepts learned in this chapter apply.

10-2 DATA ACQUISITION AS A COMPUTER PERIPHERAL

The other broad classification of the data acquisition system is concerned with its use as a peripheral to a digital computer. In this instance, timing and control are derived from the computer, and the output buffer becomes a peripheral controller. In this configuration, the computer addresses the DAS over its address bus, commands a channel to be converted, and retrieves the data when the conversion has been completed.

A system of this class may be seen in Figure 10-23. For this example, an analog multiplexer having 128 channels is utilized (channel designations 0 through 127 are used this time). A multitiered multiplexer is implemented as 8 groups of 16 switches (Chapter 5). The signal-conditioning amplifier has four gain selections that are programmed by the computer. The A/D converter has 14 bits (16,384 intervals) of resolution. The system operates in the following manner:

1. Via its application software program, the computer applies a 16-bit command or instruction word to its data bus. It also places the data acquisition system address on its address lines.
2. The computer then sets its read/write line to read, signaling the DAS to accept the instruction word on the data bus.
3. On the rising edge of the read signal, the information from the input data buffers is strobed into the instruction register. At the same time, the DAS timing is enabled.
4. The DAS interprets the command word, which is stored in the instruction register. The first 4 bits are binary-coded and indicate the analog channel selected (1 out of 16) in the first tier of switches. These bits are decoded into 16 lines before application to the switch groups. The next 8 bits are individual bits that select the switch group at the second tier. Finally, the last 4 bits determine the gain range for the signal-conditioning amplifier.
5. After time has been allowed for instruction decoding and multiplexer switch settling, the timing section issues a convert command to the A/D converter.
6. When the conversion has been completed, the timing section issues a conversion-complete signal, which sets the interrupt flip-flop, to signal the computer that the task has been completed.
7. When the computer has time, it services the interrupt. To do so, it again places the DAS address on the address bus but this time issues a write command.

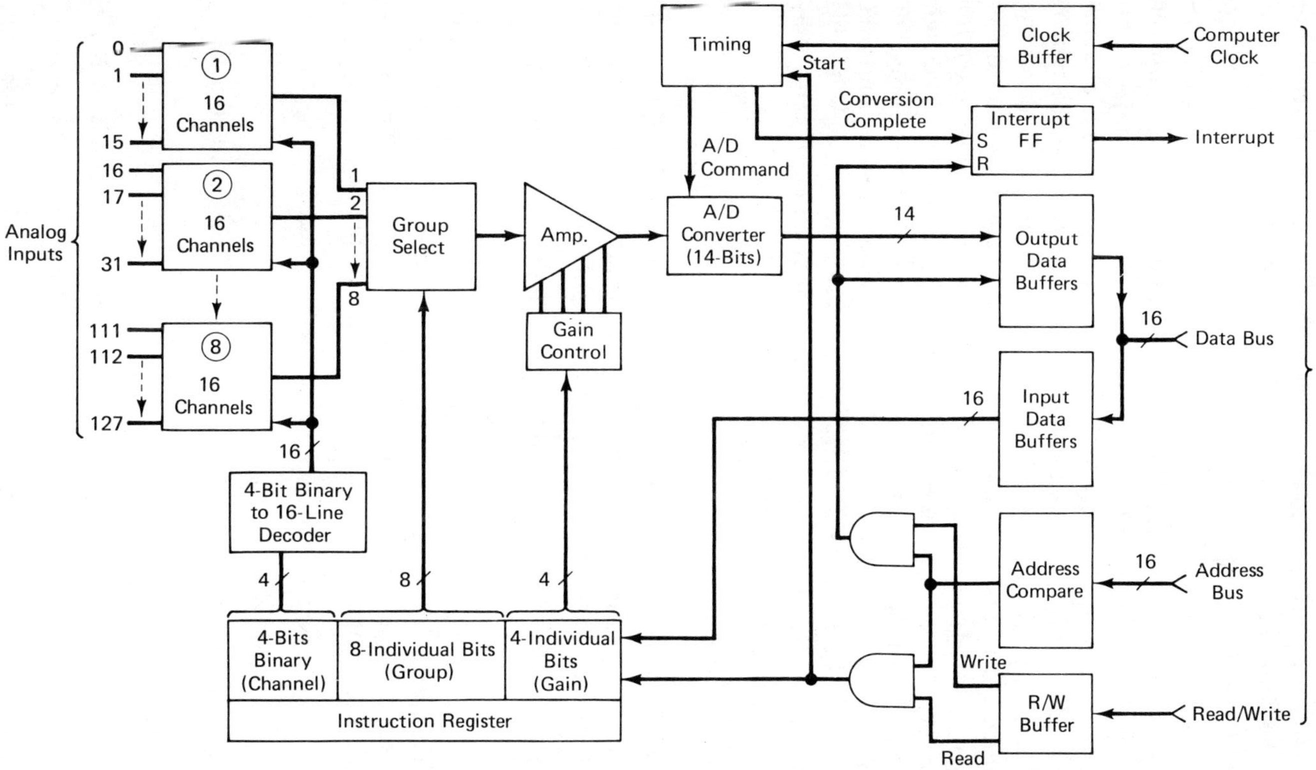

Figure 10–23. Data Acquisition as a Computer Peripheral

8. The DAS resets the interrupt flip-flop and enables the A/D converter data onto the data bus, through the output buffers. The process is now complete, and the system waits for another instruction from the computer.

As with the data logger, there are many levels of computer peripheral DAS implementations, as well as many different computer I/O architectures (Chapter 12) and a wide range of DAS complexities.

In recent years, the instrumentation industry has tried to standardize by developing a general-purpose interface bus (GPIB) system called IEEE-488. The IEEE-488 bus is a 16-line system that allows the interconnection of computers, controllers, recording devices, and test instruments. It defines a usage for each of the lines and a protocol for their use. At the time of this writing, over 1,000 instruments, vended by over 250 companies, use the standard interface and are offered for sale. In at least one case, a small microprocessor-based computer (Commodore's PET) offers the IEEE-488 bus as a standard. With this feature, many instruments and small data acquisition systems may easily be interfaced to and driven by a computer.

PROBLEMS

10–1. A data acquisition system requires 250 channels of analog multiplexing rather than the 100 of the example in Chapter 10. Develop the functional logic for such a multiplexer's control.

10–2. Develop the functional logic to provide the scan sequence pattern of Figure 10–9.

10–3. A data acquisition system has a time-of-day clock as shown in Figure 10–11. With an accuracy specification of ±0.0075 percent, what is the worst-case error in the year? Assume the drift due to temperature cancels over the year period.

10–4. A data logger has the following specifications:

1. Number of channels:	64
2. Gain steps:	3 (10 V, 100 mV, and 10 mV FS)
3. A/D resolution:	11 bits binary + sign
4. Manual data entry:	8 bits
5. Time-of-day clock:	Hours, minutes, seconds in BCD
6. End of scan:	Special 12-bit binary character

The output buffer formats all data into 12-bit words for eventual computer entry. Define a format in groups of 12-bit words for the system. In order, the data should be recorded as time, manual data, channel number, gain, sign, and data.

10–5. The data logger of Figures 10-19 and 10-20 is designed to stop at the end of the record length. What would we do if we wanted to run continuously instead of stopping?

REFERENCES

1. I. Flores, *The Logic of Computer Arithmetic*. Englewood Cliffs, N.J.: Prentice-Hall, Inc., 1963.
2. H. Hellerman, *Digital Computer System Principles*. New York: McGraw-Hill Book Company, 1967.
3. Hewlett-Packard, *An Introduction to HP Computers, 1970*.
4. Y. Chu, *Digital Computer Design Fundamentals*. New York: McGraw-Hill Book Company, 1962.
5. L. Schultz, *Digital Processing: A System Orientation*. Englewood Cliffs, N.J.: Prentice-Hall, Inc., 1963.
6. IEEE, *IEEE Standard 488–78*. 1978.
7. C. Jense, *PET and the IEEE 488 Bus (GPIB)*. Berkeley, Cal.: Osborne/McGraw-Hill Book Company, 1979.

11 Storage Units and Output Devices

11-0 INTRODUCTION

If the stand-alone data acquisition system is to be useful, it must have some type of output display or storage media. This is true even for simple systems such as digital voltmeters (DVM) or digital panel meters (DPM), whose measurements are made one at a time. With these devices, a numeric display made from light-emitting diodes (LED) or liquid crystal is used. Each time a new measurement is made, the operator reads the display and records the data. With an automatic, multichannel system, however, the observer's eyes cannot even separate the various channel inputs, let alone record them for later analysis. In this case, some type of hard-copy record or storage medium must record the data so it can be perused at human speed later or eventually processed by a digital computer.

This chapter considers some of the more common of these devices. The units discussed can be either a direct part of a basic data logging system or peripheral to a minicomputer system, of which the data logger is a part. Should the data logger's control system contain a microprocessor or microcomputer, the storage and output devices would most probably be peripheral to the computer element.

11-1 TELEPRINTERS/TYPEWRITERS

Teleprinters or typewriters are often used as output devices for low-speed data logging systems. These devices have speeds that range from 10 cps (characters per second) to 30 cps, and they provide from 72 to 80 characters per typed line. They are impact printers, utilizing hammers, print drums, or print balls to force an ink-filled ribbon against the paper to form the characters. The standard teleprinter code is the United States of America Standard Code for Information Interchange, commonly called "ASCII." The ASCII code is shown in Figure 11-1. Sixty-four characters, which consist of alphabetics, numerics, and control characters, are available. The data is transmitted to the teleprinter using the 11-bit code seen in Figure 11-2. The first bit is the "start" signal, the next 8 are the ASCII code, and the last 2 bits are the "stop" code. In teleprinter nomenclature, a *mark* is a logical one, and a *space* is a logical zero. If the teleprinter were operating at 10 cps with each character represented by 11 bits, the bit rate would be 110 bits per second (bps). It is also common to refer to a teletype "word" as equivalent to 5 characters and therefore:

10 characters per second = 2 words per second

and: 2 words per second = 120 words per minute

When you use a teleprinter, data sequencing is necessary to keep the unit from overloading. To prevent this, the data logger must get a "unit-ready" signal from the teleprinter, between characters. This signal prohibits the data logger from asking the teleprinter to print a second character, while it is still printing the first.

Some teleprinters are also equipped with a slow-speed (10-cps) paper tape punch, which can provide an 8-hole, computer-compatible tape at the same time it is typing the data. Although it is slow in comparison with the tape punches, as we will discuss later, it does provide both types of storage media.

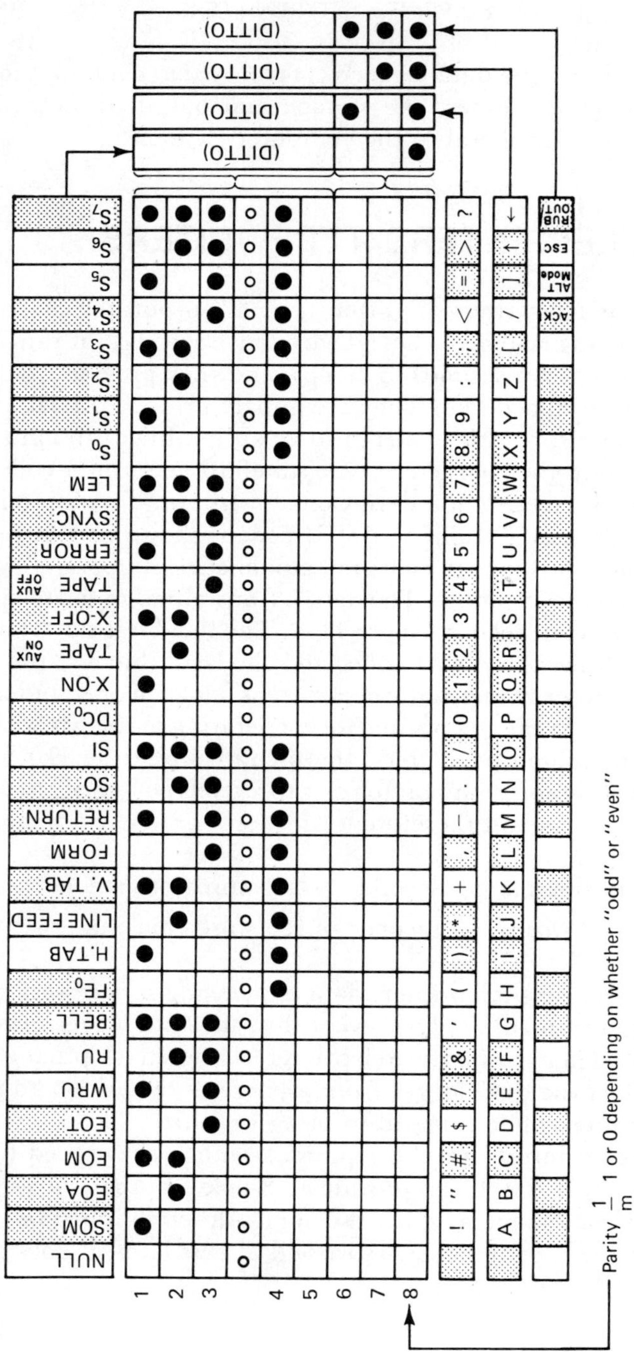

Figure 11–1. The ASCII Code

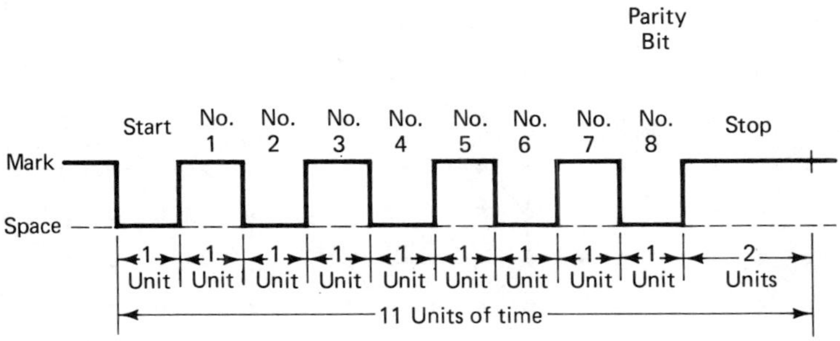

Figure 11-2. The 11-Bit Teleprinter Waveform for the ASCII Letter "U" with Even Parity

11-2 LINE PRINTERS

Although the final result for both is a hard copy output, the difference between the teleprinter and what we will call the "line printer" is speed of operation. Line printers fall into two major categories: (1) line at a time and (2) character at a time.

Line-at-a-time printers print complete 80- to 156-character lines at rates that range from several hundred to several thousand lines per minute. Since such units are usually peripheral to fairly elaborate large computer and minicomputer systems, rather than to a typical data logger, we will limit our discussion to the lower-cost character-at-a-time devices.

The most popular *character-at-a-time printer,* called the matrix printer, creates each character within a 5 × 7, 7 × 9, or 9 × 9 matrix of tiny dots. For instance, the 5 × 7 matrix printer generates each character with a print head that consists of seven vertically oriented, solenoid-driven pins. As the print head moves across the paper, a selected number of pins are exercised at five successive intervals to form a 5 × 7 array for each character. The pins impact the paper through an ink-filled ribbon to permanently mark the paper. The three characters 1, 6, and B are shown in Figure 11-3. The potential-dot circles (the ones not filled in) and the lines connecting the "printed" dots do not actually appear in the printed output, but they are shown to provide character position reference within the matrix and to overcome the effect of expanding the character size. When the characters are reduced to actual size, the dots merge to obvious, readable forms. Dot matrix printers typically range from 125 to 250 cps. Many units print both forward and backward, providing anywhere from 80 to 156 characters or columns per line. Like the slower-speed teleprinters, the signaling code used is ASCII. Again, the data acquisition system must

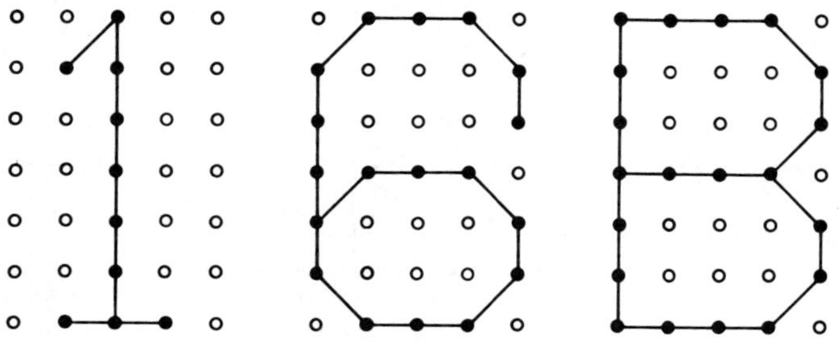

Figure 11-3. Example 5 × 7, Dot-Matrix Characters

be interlocked to the printer to prevent overloading. In this regard, the data logger waits for a "printer-ready" signal before sending the next character.

11-3 VIDEO TERMINALS

The video terminal can also be used as an output device for a data acquisition system. If the DAS contains a microprocessor, users can make random-access channel selection, as well as system operating mode selections, from the terminal keyboard. Whether the results consist of the measurement of a single channel or of a complete scan of channels, they are displayed on the cathode-ray tube (CRT) screen.

A block diagram of a video display terminal can be found in Figure 11-4. Its important display characteristics are the size of the display area, the number of characters per line, the number of lines per display, and the number of characters that can be displayed at one time. The typical video display uses a CRT that ranges between 10 and 14 inches on the diagonal. A full-screen display consists of 1,920 character spaces on a grid consisting of 80 characters per line by 24 lines. The characters are usually formed by a dot matrix (like the teleprinter), and both 5 × 7 and 7 × 9 matrices are commonly used.

As data is gathered from the data logger, it is stored in the display's refresh memory. The CRT screen is updated from the refresh memory at a rate of 60 times per second, ensuring a constant, flickerproof display. Some displays contain refresh memories with the capacity to store up to four screens or pages worth of data. Depending on the unit's sophistication, the pages can be called up one at a time or rolled up and down across the screen.

Display terminals come with both serial interfaces (usually EIA RS-232C) and parallel interfaces, and transmission rates vary from

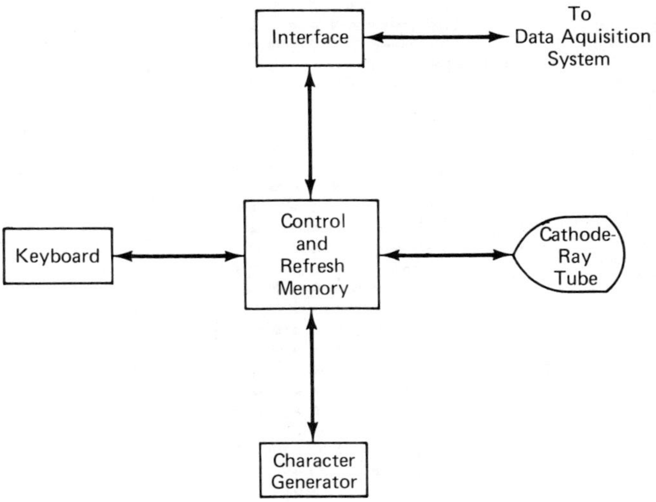

Figure 11-4. Block Diagram of Video Display Terminal

teletype speed (110 bps) to 9,600 bps. The time to transmit a 1,920 character screen's worth of data ranges from about three minutes for the teletype rate to about 2 seconds for 9,600 bps. The slower-speed units typically utilize asynchronous data transmission techniques and require start and stop bits, just like a teleprinter. The high-speed units are often synchronous and require proper control and timing signals from the data logger.

11-4 PAPER TAPE

Punched paper tape is frequently used as a recording media for slow-to medium-speed data acquisition systems. Paper tape is economical and directly computer-compatible. Data is recorded in binary or BCD on a one-inch paper tape by punching a hole to represent a logical one, or by leaving a blank (no hole) to represent a logical zero. There are eight possible bit locations across the tape, and data may be recorded in one of a number of codes. The most commonly used are the ASCII code (Figure 11-1) and the IBM code (Figure 11-5). Close observation of each representative tape shows a train of small locater holes, which are used to properly position the tape with respect to the punches. The CHECK-bit in the IBM format provides an ODD-parity check for each of the possible characters. ODD-parity requires that each legitimate character contain an odd number of holes (or "one's").

Both these codes are alphanumeric and represent numbers in BCD. Where only numbers and special characters are needed, it is not

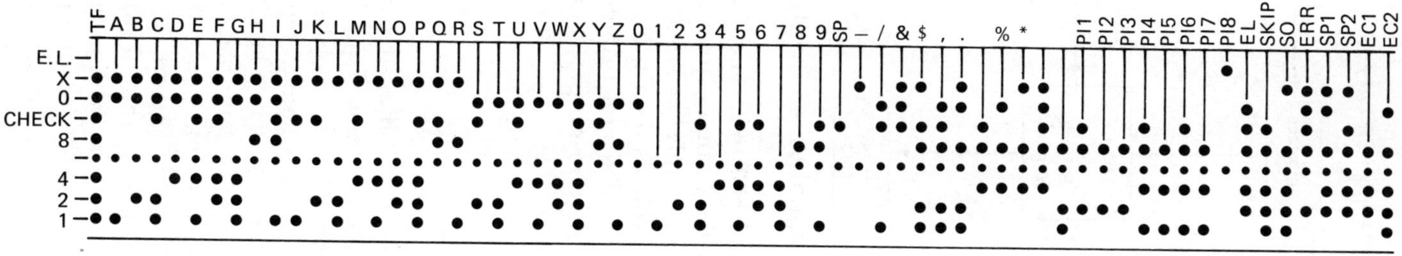

Figure 11-5. IBM Standard Punched Paper Tape Code

uncommon to utilize the eight-hole positions for binary recording. If the data words are eight bits or less, they may be represented by one tape character. Should more than eight bits be required, the data word can be represented by more than one tape character. One bit in each character can be designated as a parity bit and parity can be either odd or even.

The typical paper tape punch can record at a 50- to 60-cps rate, although units having rates of up to 600 cps may be found. The recording density is 10 characters per inch. Paper tape recorders have one other feature that is of importance to data acquisition systems. The units can be used in a start/stop or incremental mode. Often slow-speed data loggers have uneven recording rates, or they have substantial periods between samples. In this mode, the tape punch is stepped at the data logger rate, starting and stopping each time a character is recorded. This provides successive characters to be stored evenly spaced on the tape in spite of the uneven data input rate. So data entry into the data analysis computer is smooth and even. Start/stop or incremental recording is limited to rates below 300 cps.

The tapes are read in two ways: either by eight metal brushes, which make contact with a voltage-carrying conductor each time a hole passes, or by a light source, which strikes a light-sensitive detector each time a hole passes. Standard brush readers range from 60 to 300 cps, while optical units range as high as 1,000 cps.

Special signals provided by a typical punch or reader are "end-of-tape" and/or "loss-of-tension."

11-5 MAGNETIC TAPE

Three basic types of magnetic tape recorders are of interest to users of data logging equipment:

1. cassette recorders,
2. reel-to-reel recorders, and
3. cartridge recorders.

11-5.1 Cassette Recorders

The digital cassette recorder offers the data acquisition user with a higher-speed, lower-cost alternative to punched paper tape. Although many personal home computers utilize standard audio cassettes, the digital cassette system has different characteristics. First, the digital

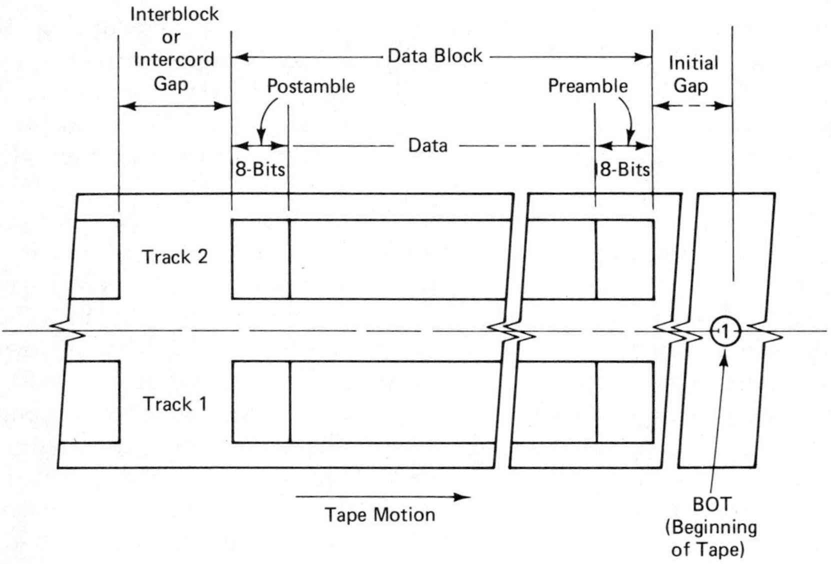

Figure 11-6. Cassette Recording Format

cassette does not need the high-fidelity characteristics of a good audio cassette. The data to be recorded consists of binary bits that are usually recorded by the presence of magnetic field (binary 1) or its absence (binary 0). As data blocks do not often come evenly spaced back-to-back, the digital recorder must accommodate incremental or start/stop operation. For efficient use of time when reading data, the digital system speeds up when it transitions the gaps between the stored data blocks.

The cassette system has some significant advantages over punched paper tape. First the cassette unit offers both recording and playback in the same package, whereas the tape punch and reader are totally separate items. Second, the cassette tape comes in a cartridge and is not as easily damaged as the paper tape. A third advantage is recording speed. The synchronous, continuous cassette can record at up to 1,500 cps, while the paper tape punch typically operates at 60 to 180 cps.

The cassette recording format can be seen in Figure 11-6. The data is recorded bit-serially, with characters end-to-end in a block. While recording or reading data, the tape moves at 15 inches per second (ips). Considering a storage density of 800 bits per inch (bpi), this constitutes a store or read rate of 12,000 bits per second (bps). During the interblock or interrecord gap, the tape reaches up to 100 inches per second. The gap allows for the acceleration and deceleration needed when going between the two rates. Note that Figure 11-6 indicates that the recorder has two record heads and that two tracks are written. It is very common to write the same data onto both tracks to improve recording reliability. Preamble and postamble bits are written at the start and end of each data block, respectively. These 8-bit groups, which are patterns of alternating one's and zero's, allow the recorder to synchronize to the data rate during data recovery.

Synchronous, continuous recorders are capable of storing as many as 700,000 characters on a tape that is 300 feet long, depending on the size of data blocks and the number of interblock or record gaps used. Incremental recorders, which start and stop with every 8-bit character, typically use much lower data rates than the continuous recorders. Their transfer rates range from 10 to 400 cps, and their storage capacity is in the 50,000- to 100,000-characters-per-tape range. Cassette recordings are highly portable, being contained in a 2.5 × 4 × 0.3-inch cartridge.

11-5.2 Reel-to-Reel Recorders

Reel-to-reel magnetic tape recorders are much more sophisticated than cassette recorders. Utilizing either seven or nine recording tracks, reel-to-reel recorders are capable of supporting up to 120,000-cps recording rates. Rather than using the presence or absence of a magnetic field to record, like the cassette, the reel-to-reel system uses a reversal of the magnetic field.

One of the most commonly used techniques is nonreturn-to-zero change at 1 (NRZ1) as shown in Figure 11-7. Here the direction of magnetic flux is changed each time a one is to be recorded, with no flux reversal when a zero is written. This scheme simplifies detection circuitry and increases recording reliability.

With the cassette system, data was recorded in a bit-serial format, while reel-to-reel systems utilize seven or nine tracks, as previously stated. In this manner, an entire BCD character, or in the case of binary data, an entire byte (8 bits) can be recorded simultaneously. It is not uncommon, therefore, to designate the abbreviation "bpi" as bytes per inch rather than bits per inch. When referring to a single

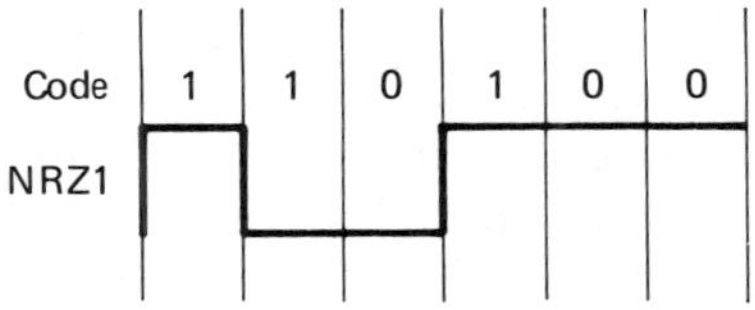

Figure 11-7. NRZ1 Recording

track, the phrase "bits per inch" is still valid.

The seven-track IBM format can be seen in Figure 11-8. As the figure shows, four different types of gaps of varying lengths are used to provide information on the position of data during the read operation. The initial gap, which occurs after the load point marker, indicates the start of the first record on the tape. The interrecord gap serves the same function for the reel-to-reel system as it does for the cassette recorder. Besides separating records, it allows a speed change during the gap. Whenever a data block or file is constructed with a series of separable records, an end-of-file gap (EOF) is used to indicate the end of the block. The LRCC gap requires a little more explanation. To start with, the P-track (Figure 10-8) is a parity bit track, which allows the system to do an error check horizontally across the character. Should two errors occur, however, the check would be fooled, and an erroneous character would be accepted as good.

The LRCC (longitudinal redundancy check character) provides another level of checking. A flip-flop, which changes state each time it sees a one, is added to each record track. At the end of record, should the track contain an odd number of ones, the flip-flop is in the one state. Should an even number of one's occur, the flip-flop ends up in the zero state. After the record has been recorded, the seven flip-flop outputs are recorded as the LRCC. When reading the tape, the flip-flops regenerate the LRCC from the data on the tape. When the LRCC is read, it is compared with the regenerated character.

Generation of an LRCC is shown for a four-character record in Figure 11-9. The seven-track convention of even parity is used in this sample. With the use of both parity and LRCC, data has been checked both horizontally and vertically. The use of even parity, as a convention, brings with it one other problem. The BCD character for zero is all zero's and, with even parity, the parity bit would also be a zero. This situation could lead to an all-zero's character and could be mistaken for part of a gap if it should occur at the start of a record. To overcome this problem, the binary number 10 (1010) is used in place of zero.

The IBM standard code for seven-track recording of BCD data can be seen in Figure 11-10. Packing densities for seven-track magnetic tape systems are 200, 556, and 800 bpi. While recording rates for continuous units range from 45 ips (9,000 bps) to 120 ips (96,000 bps), incremental, or start/stop, systems usually range from 1,200 to 5,000 bps maximum.

The standard IBM format for nine-track magnetic tape systems can be seen in Figure 11-11. This format, known as EBCDIC (*E*xtended *BCD* *I*nterchange *C*ode), has a number of advantages over the seven-track system. First, odd parity is used, and the use of the zero data

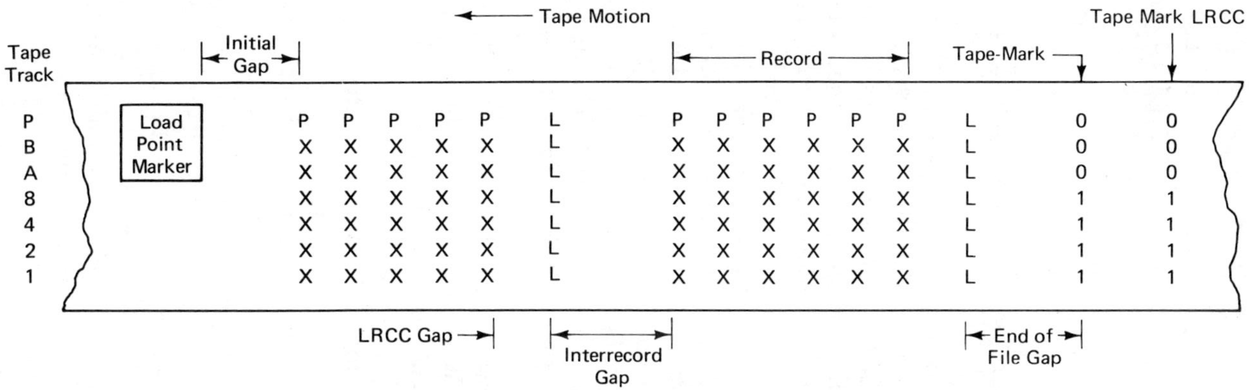

X . . . . Data Bits
P . . . . Lateral Parity Bits
L . . . . Longitudinal Redundancy Check Character

Initial Gap . . . . 3.75″
LRRCC Gap . . . . 4 Byte Spaces
Interrecord Gap . . . . .0.75″
End of File Gap . . . . .3.75″

Figure 11-8. IBM Seven-Track Magnetic Tape Format

Tape Track → 1	1	2	4	8	A	B	P
	1	0	1	1	1	0	0
	1	1	1	0	1	1	1
	0	1	1	1	1	0	0
	1	0	0	0	1	1	1
LRCC	1	0	1	0	0	0	0

Figure 11-9. LRCC Generation for Seven-Track Tape

character is no longer a problem because at least one "1" is always present. Second, notice that the lower-order bits (1, 2, 3) and the parity bit are located at the center of the tape. This places the bits with the highest rate of change in the portion of the tape that is least likely to change or flex, resulting in much more reliable recording.

Another level of error checking is also provided with nine-track systems; as we can see, another check character, the CRCC (cyclic redundancy check character) is added to the tape format. We stated earlier that the parity concept had a flaw in that an even number of errors in a character would not be detected. The LRCC also has the same flaw in that an even number of errors in a track will escape detection. To overcome these problems, the CRCC is added. Where the LRCC takes a binary count of all the 1 bits in each track, the CRCC takes a binary count of the bits diagonally through the record. CRCC character generation is shown for 4 bits in Figure 11-12 (9 bits would be too difficult to picture). With all three error checks (parity, LRCC, and CRCC), it is very difficult to read an erroneous record.

Nine-track continuous recorders have recording densities that range from 800 to 1,600 bpi and recording rates from 96,000 to 240,000 bps.

Very often users of either seven- or nine-track recorders would not like to have their computer analyze every record on the tape. They would prefer to use a high-speed search mode to select a specific record. The basic circuitry to do so is found in Figure 11-13. When the data is recorded, the record address is stored in the first character location of each block. When the computer initializes the tape unit, it stores the address of the desired block in the unit's address selection register. When in search mode, the tape unit examines the address characters located immediately after the gaps and preambles. When the correct address is detected, the address comparator enables the address-found signal, and the computer reads the record or block. This discussion is, of course, a simplification. In actual practice, the tape unit moves from block to block, stopping, reversing, and moving forward again.

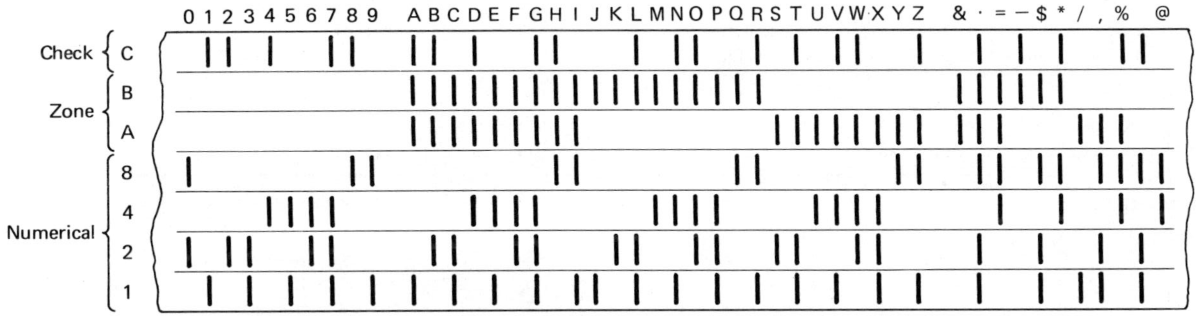

Figure 11-10. IBM Standard Seven-Track Code

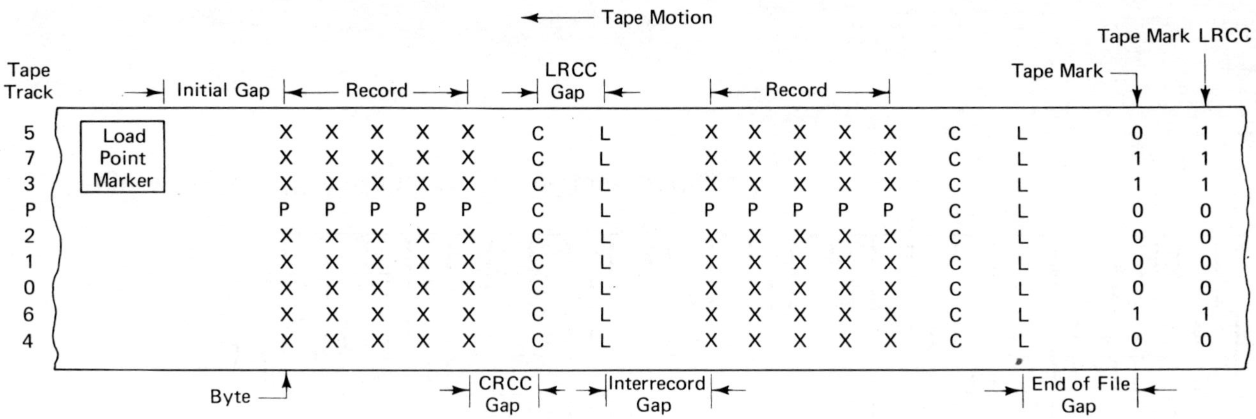

X .... Data Bits
P .... Odd Lateral Parity Bits
C .... Cyclic Redundancy Check Character (CRCC)
L .... Longitudinal Redundancy Check Character (LRCC)
Record .... 12 Characters minimum

Initial Gap .... 0.5″ min. (3.75″ typical)
CRCC Gap .... 4 Byte Spaces
LRCC Gap .... 4 Byte Spaces
Interrecord Gap .... 0.6″, + 0.15, − 0.10″
End of File .... 3.75″

Figure 11-11. IBM Standard (EBCDIC) Nine-Track Magnetic Tape Format

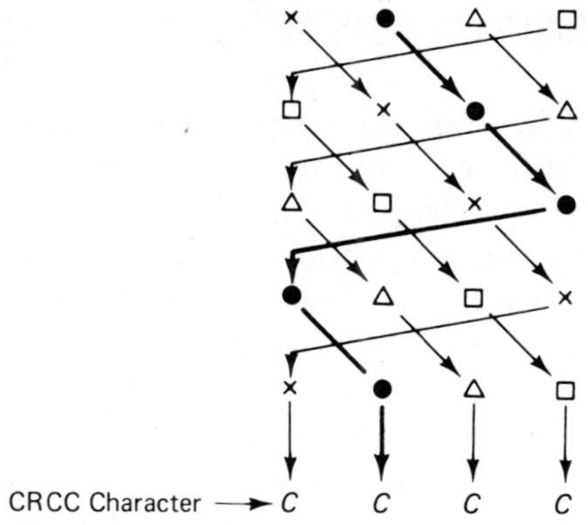

Figure 11-12. CRCC Character Generation

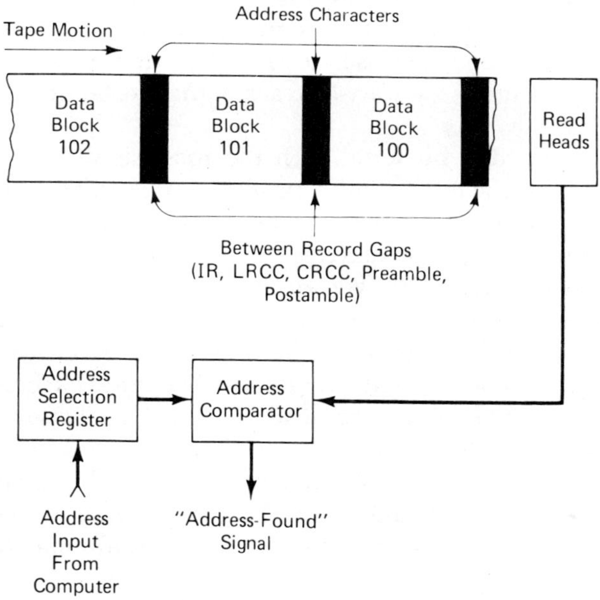

Figure 11-13. Record Location Circuitry

11-5.3 Cartridge Recorders

The cartridge recorder is an attempt to combine reel-to-reel performance with the easily handled tape cassette. A typical unit, such as the 3M cartridge, can read and write at 90 ips and search at 180 ips, with a recording density of 1,600 bpi (144,000 bps). Like the reel-to-reel system, data is recorded in multiple tracks. The drawback, of course, is that the tape length is much less than the reel-to-reel system.

11-6 THE FLOPPY DISK RECORDER

The floppy disk is the most commonly used high-speed recording device for microprocessor and microcomputer systems. It is also widely used with the smaller, lower-cost minicomputer systems. For this reason, it is often found as an output device for a data logger. Should the logger contain a microprocessor, interface hardware is usually available. The floppy disk storage medium, called a diskette, can be seen in Figure 11-14. The diskette, which looks much like a 45-rpm phonograph record, is constructed from a thin flexible mylar® substrate coated with a magnetic oxide—hence, the name *floppy* disk. Seventy-seven concentric recording tracks are contained on the diskette, with track 00 located at the outer edge of the disk and track 76 located near the center. The movable read/record head may be positioned over any one of the 77 tracks on command. The start of a track is indicated by the index access hole. The diskette rotates at a 360-rpm rate, causing an index or start-of-track signal to be generated every 166.67 msec.

Data is recorded bit-serial with the most significant bit (MSB) recorded first and the least significant bit (LSB) last. The typical read/write rate for a floppy disk system is 250,000 bits per second, or 31,250 bytes per second. A single diskette has the capacity to store over 2 million bits or over 250,000 bytes on its 77 tracks.

A primary reason for using the floppy disk is its excellent access time. Whereas a tape system would have to search the entire tape if the desired record were at the end, the floppy disk simply accesses the last track by moving the read/record head directly to it. Time to move the head from one track to the next is typically 10 msec per track, and the settling time, once the correct track has been reached, is 10 msec. Considering that the data required is, on the average, half-way around the disk from start of track, the latency is 83.3 msec (half of a 166.67-msec revolution).

Two types of recording formats are used with the floppy disk system. In either case, they call for the disk to be divided into segments or sectors. With the *index recording* format, the sectors do not define physical position on the disk but rather identify the boundaries of the data records. This arrangement allows users to write either one long record or a series of variable length records. Each track is identified by an index address mark, and each data block or record is preceded by a unique sector address identifier.

The *hard sector recording* format requires that the disk be physically separated into 26 or 32 sectors (depending on the system). Each

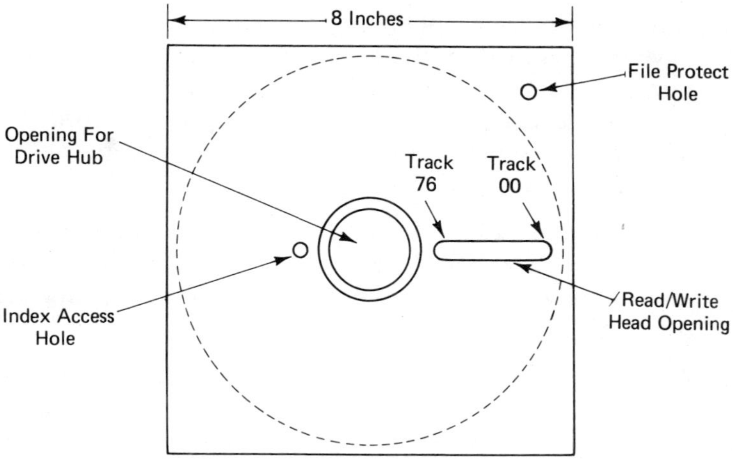

Figure 11-14. Floppy Disk Diskette

track is identified by a physical index pulse, and each sector is identified by a physical sector pulse. So the physical data position on the disk is precisely established. Hard sectoring or soft sectoring is established by the floppy disk drive and is not selectable by the user. Data format within the sectoring is at the choice of the user and highly flexible.

To protect the data, two identical cyclic redundancy check (CRC) characters are utilized. These characters are developed by counting all the bits in a record (from bit 0 of the address mark to bit 7 of the last byte) and recording it at the end of the record. Two redundant circuits accomplish this and two identical characters are written, one after the other. When the record is read, the same circuitry regenerates the characters and divides them by the CRC characters, which were stored. If there is no remainder, the record is good. Should there be a remainder, the record is suspect.

PROBLEMS

11-1. A data acquisition system has 100 multiplexed channels and a 4-digit A/D converter. It must record the time of day and 3 digits of manual data at the start of each record, along with an EOR digit at the end of the record. Channel number, sign, and data must be recorded for each channel in the record. How long would it take to record the results from one record if the output device were a 10-cps teleprinter?

11-2. If the recording device in problem 11-1 were 150 cps and if a paper tape punch were used, how long would it take then?

11–3. A data acquisition system has 24 channels of input data, each having a bandwidth of 100 Hz. Each channel is sampled at a rate of 5 samples/cycle and requires 3 digits of channel number, 1 digit of sign, and 4 digits of data to be recorded. What nine-track magnetic tape units would be acceptable for this system?

11–4. If the system of problem 11-3 used an incremental tape recorder with a record rate of 5 KHz, what would the maximum bandwidth of each channel be, assuming they were all the same?

11–5. A data acquisition system uses a nine-track, 800-bpi recorder. Each day it must store 20 records of 1,000 8-bit words per record. An IR gap is placed after each record, along with both an LRCC and CRCC (Figure 11-11). An end-of-file character is written at the end of each day's recording.
a. How much tape is used per day?
b. If the tape is 2,400 feet long, how many days will it last?

REFERENCES

1. J. D. Lenk, *Handbook of Microprocessors, Microcomputers, and Minicomputers*. Englewood Cliffs, N.J.: Prentice-Hall, Inc., 1979.
2. C. Weitzman, *Minicomputer Systems*. Englewood Cliffs, N.J.: Prentice-Hall, Inc., 1974.
3. I. Flores, *Peripheral Devices*. Englewood Cliffs, N.J.: Prentice-Hall, Inc., 1979.
4. G. Korn, *Minicomputers for Scientists and Engineers*. New York: McGraw-Hill Book Company, 1973.
5. J. Taunt, *Digital Tape Drives*. Elmhurst, Ill.: Business Press, 1965.
6. D. Reichel, "Flexible Discs: A Look at the Latest Mini-Peripheral," *EDN* (January 20, 1975).

12

The Digital Computer

12-0 INTRODUCTION

Perhaps no other single factor has impacted the data acquisition field as much as the evolution of the digital computer. High computational speed, efficiency, and flexibility have greatly increased the desire to translate information into a format that the computer understands.

Several years ago, computer manipulation of such data was reserved for very special situations. With the introduction of small computers, such as the mini- and microcomputer, along with the subsequent low cost of computing hardware, even the small low-budget experimenter can now utilize the computer for data acquisition. Such being the case, a book on data acquisition systems would not be complete without a chapter on the digital computer.

This chapter is designed to introduce you to the very basics of the digital computer, as well as to provide an understanding of the terminology. It does not pretend to be an in-depth study of the subject. Many excellent texts on the topic of digital computers are available, and reference to some is found at the back of this chapter.

12-1 THE BASIC COMPUTER

The digital computer is composed of five basic functions, as can be seen in Figure 12-1. A brief description of the major functions are as follows:

Arithmetic Unit

The arithmetic unit is at the center of the digital computer system. The unit does just what its name implies: It performs arithmetic calculations and manipulates the data. The unit contains logic that combines the contents of two registers, or two parallel-word inputs, to perform addition, subtraction, multiplication, division, data management, and formatting, as well as comparison and logical operations.

Memory

The memory referred to in the block diagram of Figure 12-1 is the computer's main memory. It contains the program or set of instructions that the computer is to execute, as well as the data the program is to operate on. Each item or word of digital data stored is assigned to a unique physical location or address in the memory unit. When the computer wishes to store or retrieve data, it simply supplies the desired address to the memory unit. The process of retrieving data from the memory unit does not destroy or erase the data in memory. In effect, the arithmetic unit simply copies the data. The speed of most computers is limited by the speed of the memory unit.

Control

The control section provides all the timing and routing information needed to operate the computer. It interprets the instructions read from memory to steer data throughout the computer and to set up the function of the arithmetic unit.

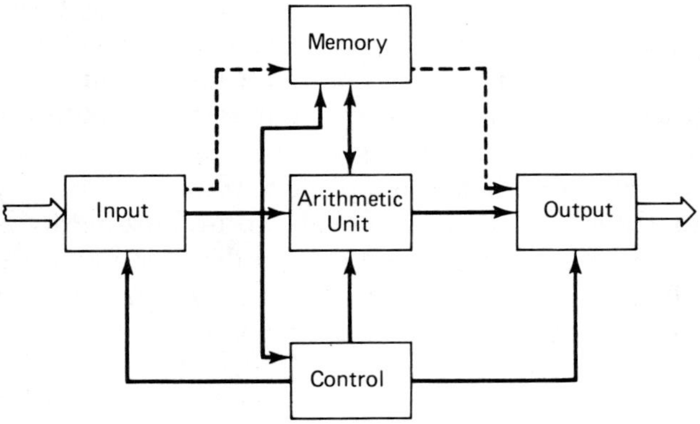

Figure 12-1. Basic Block Diagram of the Digital Computer

Input and Output Sections

The computer is useless unless it can communicate with the outside world. Data to be processed must be fed into the computer, and the results of the processing provided to the user. This activity is accomplished through the computer's input and output sections. In most cases, input and output are combined to form the I/O section.

In the following sections, we take a more in-depth look at the operation of the digital computer. Keep in mind one very important concept as we proceed: The computer is a very fast "dummy." It does precisely what we tell it to do (right or wrong) at a very high speed.

12-2 THE STORED-PROGRAM CONCEPT

Early digital computers were based on fixed, or hard-wired, programming. In other words, the computer had one basic program that it could execute and that was permanently wired in. To change the program, the machine would have to be disassembled and rewired in a different way. Only data was stored in a changeable medium. Such was the fixed-program technique.

Today's computers utilize what is known as the stored-program concept. This concept allows for the modification of data, instructions, and addresses. In the stored-program computer, both program instructions and data are stored in changeable, read/write main memory, as seen in Figure 12-2. The computer works on a concept of alternate instruction fetch and execute. First an instruction is read from memory. After interpretation, the data paths are established, and the arithmetic unit is conditioned for the required operation. With this accomplished, the operation is carried out.

This machine concept offers a high degree of flexibility. As the program is stored in main memory, it is easy to change. The computer does not have to be rewired. Most often the computer executes the instructions in sequence, as they are stored in memory. Yet with the instructions in main memory, branching may occur as the result of computation; that is, if certain conditions are satisfied during the previous calculation, the program may jump to an instruction address that is out of sequence. Looping may also occur, which is useful when the same group of instructions is used over and over. At the conclusion of a pass through the instruction group, the computer may jump back to the first instruction, depending on the results of the previous pass.

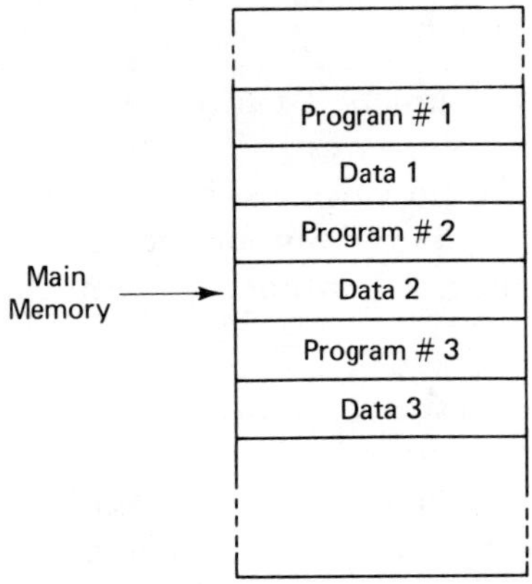

Figure 12-2. Stored-Program Computer Memory

12-3 SINGLE-ADDRESS COMPUTERS

Many minicomputers and microcomputers available today are single-address computers. That is, each instruction contains one operation code and one operand address. This type of instruction format is shown in Figure 12-3a. Although they will not be discussed in detail here, there are some architectures which utilize multiple-address instruction words. These architectures are particularly valuable where memories are slow and speed is important, as it allows retrieval of more than one operand at a time. A typical double-address instruction can be seen in Figure 12-3b.

Some older computers utilized as many as four operand addresses in their instruction word. As architectures became more sophisticated and hardware became faster, the minicomputer evolved to the simpler one-address machine. The search for speed continues, however, and, as speed surpasses hardware capability, computer manufacturers are turning again to multiple-address architectures. For the purposes of this text, our discussion will be confined to single-address machines.

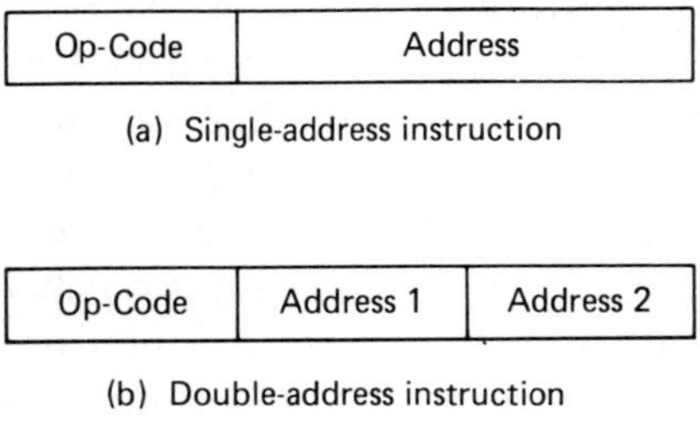

Figure 12-3. Single-Address Versus Double-Address Instructions

The block diagram of the basic single-address computer can be seen in Figure 12-4. Although still a fairly simple structure, we find a few blocks in this diagram that did not appear in the basic block diagram of Figure 12-1. Before proceeding, let's describe the operation of these new boxes.

Program Counter

The program counter (PC) contains the address of the next instruction to be executed. Since the program counter contents always pass through the arithmetic unit before being applied to the memory to fetch the next instruction, it can be readily modified to accommodate such things as branching and looping.

Memory Address Register

The memory address register (MAR) contains the address of the memory location currently being accessed. For a data or instruction word to be read from memory, the address is placed in the memory address register, and the read/write line set to read. To store a word, the address is placed in the MAR and the read/write line set to write.

Memory Buffer Register

The memory buffer register (MBR) contains the data or instruction word currently being read from or stored into memory. If the read/write line is set to read, the MBR contains a word being retrieved from memory. Should the read/write line be in the write position, it contains the word being stored.

Instruction Register

The instruction register (IR) contains the instruction currently being executed. Usually the operand address portion of the instruction is transferred directly from the MBR through the arithmetic unit to the MAR to fetch the operand, while only the op-code portion of the instruction (Figure 12-3a) is stored in the IR.

Accumulator Register

The accumulator register (AR) contains one of the operands to be operated on during the execution of an instruction. The other operand is held in the memory buffer register (MBR). After the operation has been completed, the result is stored back into the accumulator. In the architectures we will investigate in later sections, the single accumulator is often replaced with a bank of general-purpose registers. Then

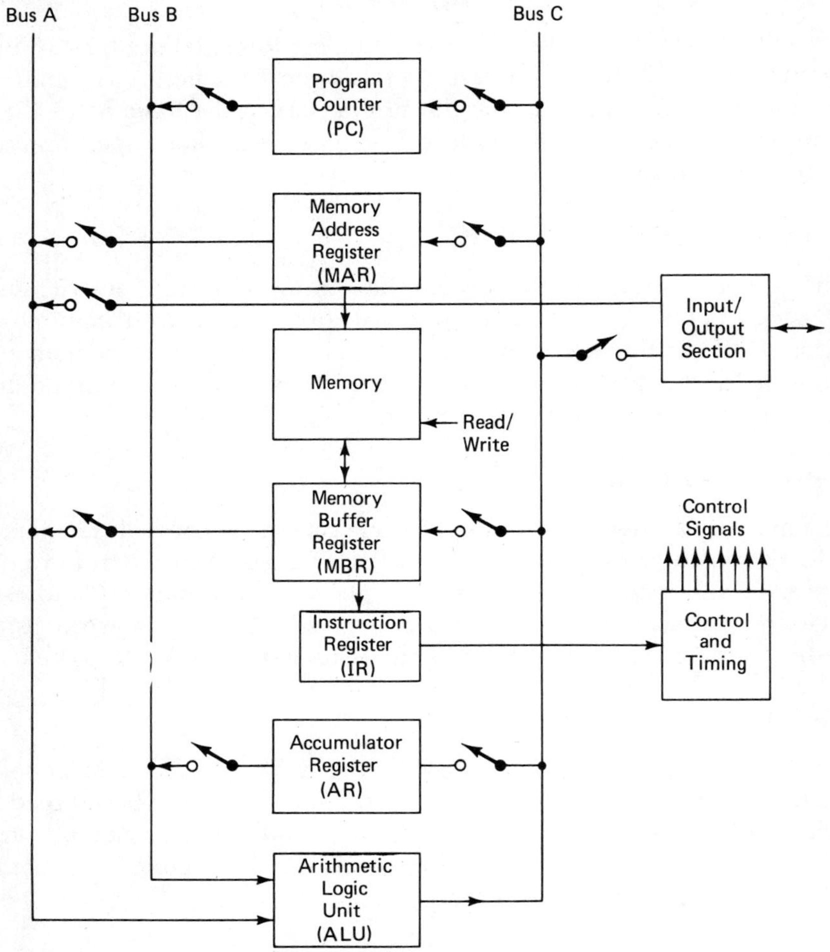

Figure 12-4. Basic Single-Address Computer

an operand in one register can be operated on by an operand in another, and the result of the operation stored in still a third register.

Our simple computer contains three buses—two input (A and B) and an output (C). Let's examine some simple operations that could occur in our computer. To do so, we have to learn two symbols or conventions:

$\rightarrow$ = shows the direction of data transfer

() = the contents of ______

As a first example, let's simply take a data word from memory and store it in the accumulator. The operations proceed as follows:

1. (PC)→MAR

   The contents of the program counter are loaded into the memory address register. To do so, the switch from the program counter to bus B and the switch from bus C to the MAR must be closed. The address passes through the arithmetic unit to the MBR.

2. INSTRUCTION→MBR

   The contents of the address pointed to by the MAR are stored in the MBR. The instruction has been "fetched" from memory.

3. OP-CODE→IR
4. ADDRESS→MAR

   The instruction is split: The op-code portion is stored in the instruction register, while the operand address is routed to the MAR via buses A and C and the arithmetic unit.

5. OPERAND→MBR

   The contents of the data address pointed to by the MAR are stored in the MBR.

6. (MBR)→AR

   The instruction is executed, as the operand is stored in the accumulator.

As a second example, let's add a number to the one we just stored in the accumulator:

1. (PC)→MAR
2. INSTRUCTION→MBR
3. OP-CODE→IR
4. ADDRESS→MAR
5. OPERAND→MBR
6. (MBR) plus (AR)→AR

Operations 1 through 5 are identical to those of the first example. Only in step 6, the instruction execution, do they differ.

The input/output (I/O) section is also connected to the computer's internal buses, and, as usual, the data is always handled by the arithmetic unit.

12-4 COMPUTER WORD LENGTH

Digital computers are often classified by their data and instruction word lengths. Our definition of word length coincides with the physical length of the data buses and main arithmetic registers in the computer. With this point in mind, let's look at some of the available word length computers.

8-Bit Word Length

Up to the end of the seventies, 8-bit machines were considered to have limited capabilities as full-fledged computers. Most of those were organized as controllers rather than as computers. With such a short word length, they had limited instruction power and addressing capability, and so they were restricted to uses in which a limited number of repetitive actions were required.

The advent of the monolithic microprocessor has changed all this. Through clever architectures, multiple registers, and efficient software and firmware, these devices have provided "big machine" computing capability with basic computers of small word sizes. In many available systems, up to 12-digit arithmetic is provided under software or firmware control. The penalty is, of course, speed of computation, since the arithmetic is accomplished a digit at a time. In a great many data acquisition systems, speed in the processor is not a problem, and the personal computer works quite well.

12-Bit Word Length

The 12-bit computer offered more computer power but still suffered from limited instruction and addressing capability. Characterized by the Digital Equipment Corporation PDP-8 family of computers, this word length computer was commonly utilized in dedicated operations where single tasks are required.

16-Bit Word Length

The most common of all minicomputer word lengths is 16 bits. This word length offers an excellent balance between instruction set power and memory addressing capability for a short word length computer. Most 16-bit computers are capable of handling multifunctional tasks, and they successfully bridge the gap between dedicated systems and stand-alone multipurpose systems.

32-Bit Word Length

Relatively inexpensive integrated circuit hardware has pushed the minicomputers solidly into 32-bit word lengths in the last five years of the seventies. This provides a minicomputer with big computer power. It allows the computer to have a rich instruction repertoire, while allowing direct addressing of a substantial amount of main memory. Such machines handle multifunctional tasks with ease, and many can be utilized for time-sharing systems.

Besides the problem of instruction power versus address space in the instruction word, word length has an impact on the accuracy of arithmetic computation on the data words. The problem occurs because most minicomputers do not perform a true round-off function. They simply truncate, or chop off, any digits beyond the word length's least significant bit. Although the final number is never larger than the correct quantity, the least significant bit can easily be in error, and the computer does not provide the n-bit of accuracy implied. Double precision arithmetic, or some type of averaging would have to be used to regain the lost accuracy.

12-5 ADDRESSING MODES

As we hinted in the previous section, a limited word length can severely limit the computer's ability to couple a reasonable size instruction set with good memory addressing capability. To better understand this point, let's examine some popular ways of addressing.

Direct

When in the direct addressing mode, the computer instruction word contains the actual address of the operand. The basic instruction word format for direct addressing can be seen in Figure 12-5. The word is comprised of two parts: the operation code (op-code) field and the address field. In the typical 16-bit minicomputer, the op-code field ranges from 4 to 6 bits, while the address field ranges from 10 to 12 bits. In the best case, the instruction can address only 2^{12} or 4,096 memory locations. This severely limits the machine unless some of the addressing schemes discussed in the sections that follow are used. You can certainly see the advantage of a 32-bit word length, since the usual 32-bit computer allows at least 16 bits for direct addressing. This provides direct access to 65,536 words of memory.

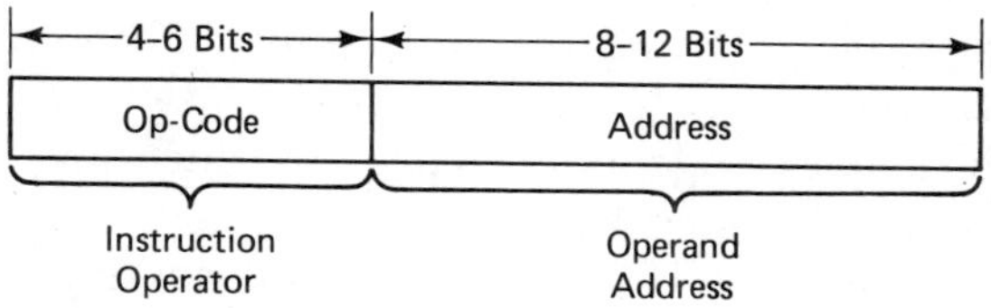

Figure 12-5. Direct Addressing Instruction Word

Indirect Addressing

To get around the problem of a restricted address field, the minicomputer turns to alternate addressing modes to achieve adequate addressing range. One of those most commonly used is that of indirect addressing.

The indirect mode, seen in Figure 12-6, utilizes the address portion of the instruction word to point to the memory location of the actual operand address. At the sacrifice of two fetches, the memory access has been increased to 16 bits. A special bit (bit 15) has been added to the instruction word to designate which mode the computer is in—direct or indirect. When in the direct mode, the instruction can address up to 1,024 locations (10 bits), while in indirect it can access up to 65,536 locations. Sometimes two levels of indirection are desired, and bit 15 indicates if the second word is also acting as a pointer.

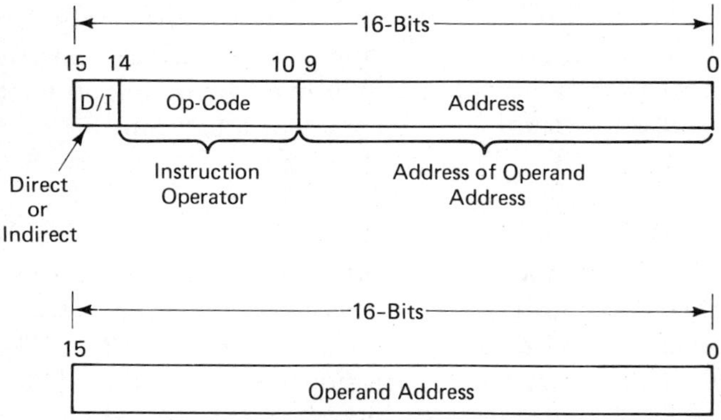

Figure 12-6. Indirect Addressing

Paging or Sectoring

Another method for increasing memory access despite limited word length is through paging or sectoring, shown in Figure 12-7. The memory is segmented into pages or sectors commensurate with a good instruction-set/address-length mix. Typically a 16-bit minicomputer page is either 512 or 1,024 words in size. As Figure 12-7a indicates, we have chosen pages of 512 words for our example. The lower-order page (page 0) is also referred to as the "base page." Users move back and forth between the base page and the page in which they are currently working simply by changing bit 9.

Suppose, for instance, that the program is working in page 2 (current page) and at location 1,151. Suppose also that the instruction fetched at location 1,151 has the B/C bit set to B and an address of 130 in the operand address field. The computer jumps to the base page

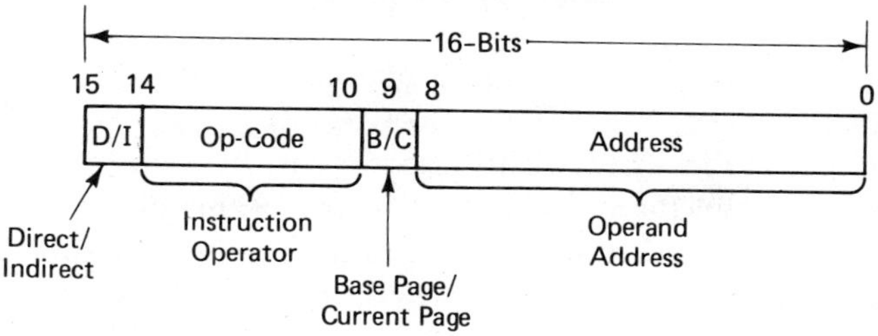

(a) Instruction word

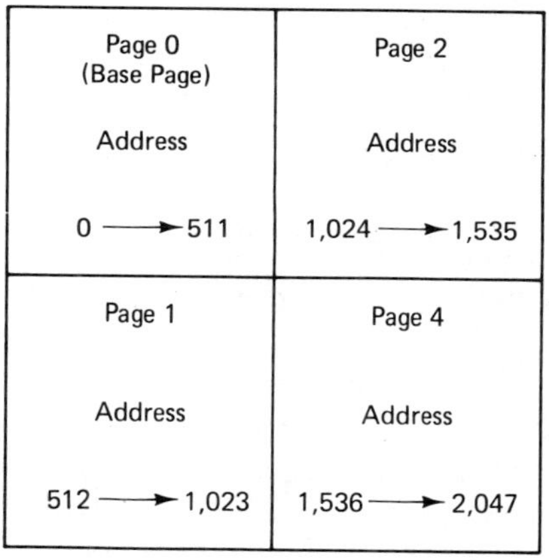

(b) Memory map

Figure 12-7. Page or Sector Addressing

and utilizes an operand located at address 130 to supplement the routine it is working on in page 2. The program counter has not been altered by this page-jumping and so the next instruction executed is taken in sequence from location 1,152. Paging can be handled either by having a special page register or as a current page extension to the program counter. In either case, the page number must be established through the software by instruction.

Index and Base Register Addressing

Indexing is another common form of adding versatility to memory addressing. In this case, the computer has one or more registers, called *index registers*. These registers may be addressed and a number inserted. This number is added to the operand address field of the current instruction to determine the actual operand address. The registers may be incremented, decremented, or tested.

The instruction word with indexing can be seen in Figure 12-8. As we see, another bit must be given up to accommodate the new mode. Each additional improvement in addressing capability reduces the available space for direet addressing or reduces the available op-codes.

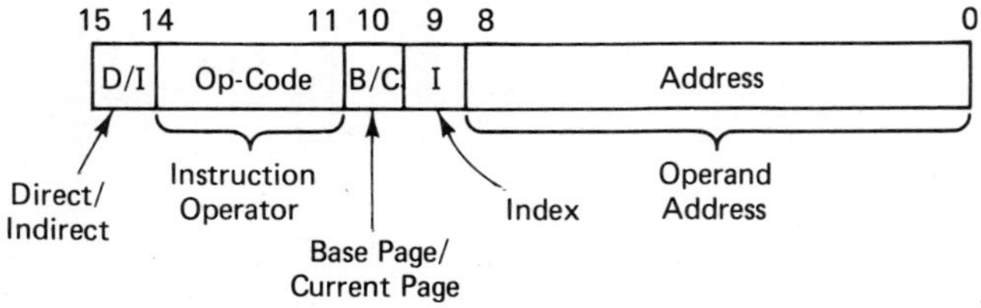

Figure 12-8. Instruction Word with Indexing Added

The indexing operation can be seen in Figure 12-9. To start, the number 1,024 is loaded into the index register. When the next instruction is fetched, the I bit (bit 9) is set to a logic 1, denoting the index mode. At this time, the operand address is added to the contents of the index register. The result, 1,261, is the actual operand address. When loaded into the memory buffer register, it causes the operand, located at that address, to be loaded into the accumulator. This would constitute the execution of the instruction, load accumulator-indexed. After the instruction execution is complete, the index register is incremented to 1,025, and the next operand is selected from address 1,262. In this way, an instruction can be made to operate on data located anywhere in memory.

Indexing is extremely powerful if you wish to move data from one part of memory to another.

Another technique that is similar to indexing is the use of a base register. In this situation, the register is loaded with a "base" number, which would be a constant and not change. This number allows you to utilize a series of instructions to operate on a block of data residing anywhere in memory. The operand locations would be displaced by the value of the base number.

Floating Pages, or Relative Addressing

Another paging scheme that is commonly used provides for a base page and a current page, which floats with its center about the current instruction. This technique, which can be seen in Figure 12-10, is

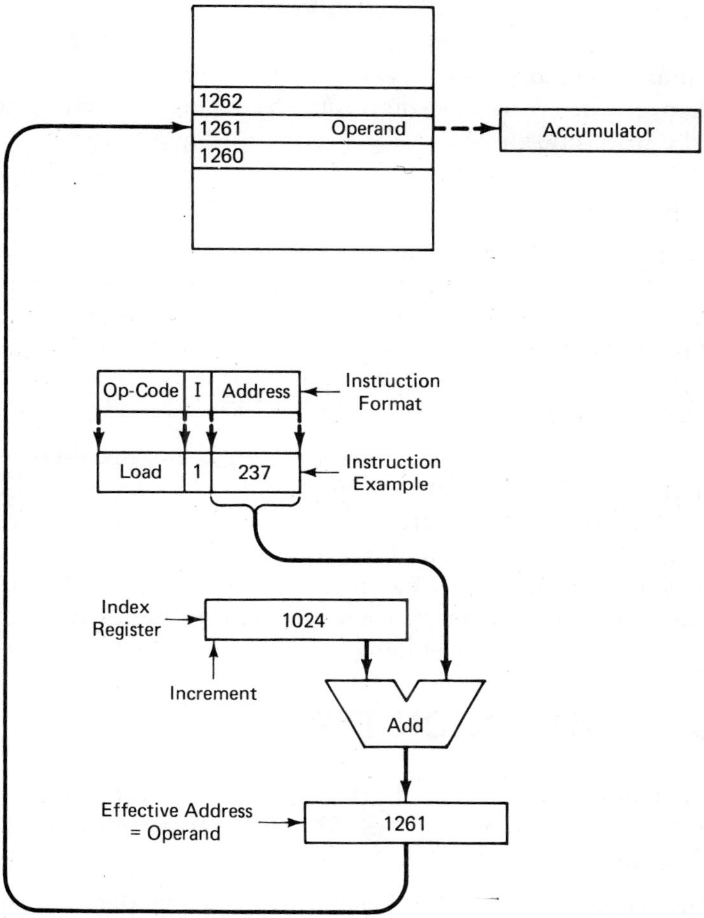

Figure 12-9. Indexing Example

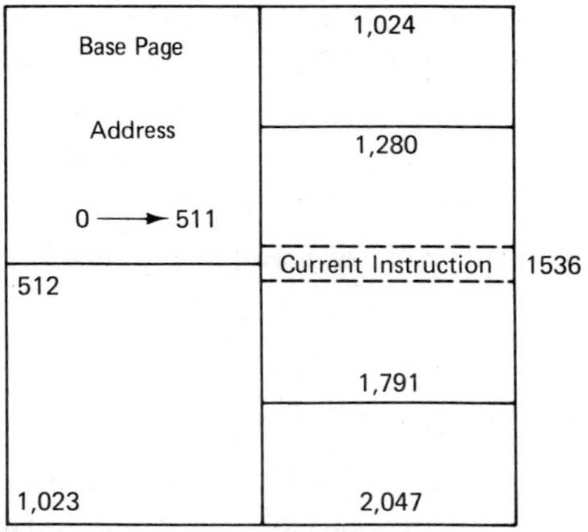

Figure 12-10. Floating Pages, or Relative Addressing

especially useful when the instructions and data to be operated on are contiguous in memory. Once again, transfer from the base to the current page and vice versa is accomplished by setting or resetting the B/C bit in the instruction word.

Immediate Addressing

In the immediate address mode, the actual integer, or operand, is contained in the operand address field portion of the instruction word. This technique is useful whenever the program contains a constant that is used often.

Bank Switching

Often a minicomputer memory is separated into banks. These banks are usually of the maximum size the computer can address under normal conditions—direct, indirect, and so on. A bank register is provided, and the computer can alter the register under program control. Bank 0 is the computer's normal memory. Other banks are extra memory added to the computer. Since a program segment cannot straddle banks, take care when developing software.

12-6 INSTRUCTION TYPES

A typical instruction set for a 16-bit minicomputer can be seen in Table 12-1. The computer has 32 basic instructions that can be broadly categorized as follows:

1. Instructions 0 through 6 load, store, or move data.
2. Instructions 7 through 15 perform arithmetic operations.
3. Instructions 16 through 29 (with the exception of 23) are data comparison instructions. The computer branches or skips, depending on the results.
4. Instructions 30 and 31 allow rotation and shifting of data in the accumulator (or data register) in a multiregister computer.

Let's now examine some of these instructions in more detail.

Load/Store

The load/store instructions move data from memory to the accumulator or to a general-purpose register, and vice versa. An example of a load accumulator instruction can be found in Figure 12-11. The computer operation is comprised of two parts: an instruction fetch cycle and an instruction execution cycle. During the fetch cycle, the pro-

TABLE 12-1

Code	*Instruction*	*Description*
0	STS	Store single precision
1	STD	Store double precision
2	LDS	Load single precision
3	LDD	Load double precision
4	LDC	Load one's complement
5	LDN	Load two's complement
6	SWP	Swap
7	ADD	Add single precision
8	ADP	Add double precision
9	SUB	Subtract single precision
10	SDP	Subtract double precision
11	AND	Logical AND
12	IOR	Logical inclusive OR
13	XOR	Logical exclusive OR
14	MPY	Multiply
15	DIV	Divide
16	JMP	Jump unconditional
17	JPZ	Jump if plus or zero
18	JNG	Jump if negative
19	JEZ	Jump if zero
20	JNZ	Jump if nonzero
21	JOF	Jump if overflow set
22	JSB	Jump to subroutine
23	EXQ	Execute
24	ISZ	Incr and skip if zero
25	DSZ	Decr and skip if zero
26	CSL	Compare and skip less
27	CSE	Compare and skip equal
28	ASZ	And and skip if zero
29	OSF	OR and skip if full one's
30	RSP	Rotate single precision
31	SAD	Shift arithmetic double precision

gram counter (PC) contents are placed in the memory address register (MAR), and the instruction obtained from the location pointed to is placed in the memory buffer register (MBR). The computer's control unit decodes the operation code (op-code) and causes the following actions: The operand address portion of the instruction is placed in the

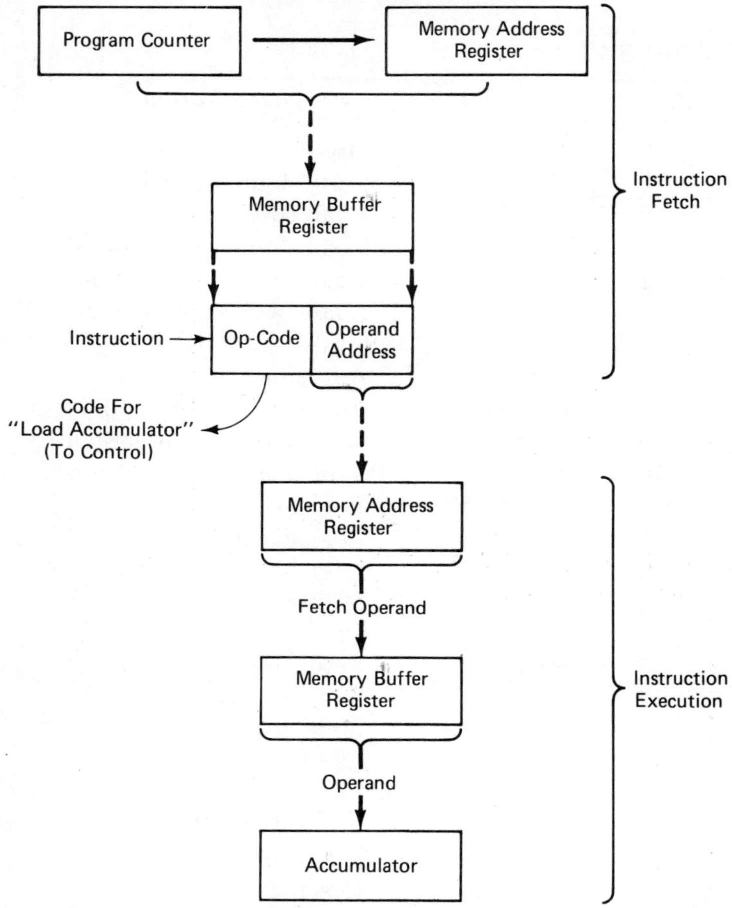

Figure 12-11. "Load" Accumulator Instruction Execution Map

MAR, and the data is fetched to the MBR. The final instruction action places the contents of the MBR in the accumulator, and the execution cycle is complete.

Had load double precision been executed, two words would have been fetched. The first would be stored in the accumulator and the second in the extension register (ER), which is concatenated with the accumulator. The higher-order bits go to the accumulator while the lower-order bits go to the ER.

Add/Subtract

The execution map for the add instruction can be found in Figure 12-12. The operation of the instruction is identical to the load instruction up to the last step. At that point, the arithmetic unit adds the operand stored in MBR to that stored in the accumulator. The result is stored back into the accumulator. The subtract operation is identical to the add, with the exception that the operand is complemented (the sign changed) before the add operation takes place.

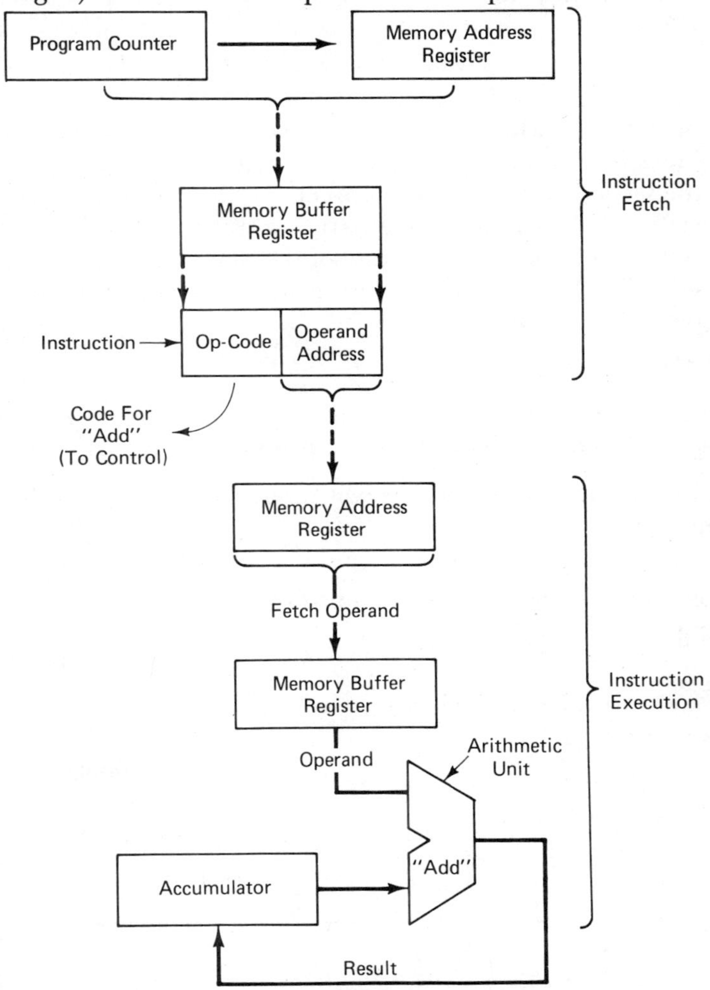

Figure 12-12. "Add" Instruction Execution Map

Overflow and Carry Flags

After an arithmetic operation, the arithmetic unit sets up special flags to indicate whether a carry has resulted or whether the result is too large for the accumulator (overflow). Carry flags are useful in positive integer arithmetic (especially one's complement) or in the case of double precision shift or rotate instructions.

The overflow mechanism can be seen in Figure 12-13. In this diagram, A = sign carry and B = most significant bit carry. Overflows occur when:

$$\text{Overflow} = A\bar{B} + A\bar{B}$$

That is, it occurs when there is an A carry and no B carry, or vice-versa. The first situation accounts for a negative number overflow, while the second is the positive number overflow. In many computers, the overflow and carry bits are a part of a status register. This register contains information as to the state of the machine after each instruction execution. What information is contained depends upon the specific computer design.

Logic Operations

There are three basic logic operations—AND, OR, and COMPARE (exclusive OR). Operational definitions are as follows:

1. AND replaces selected bits with zero's by masking

1100101	Word
0000100	Mask
0000100	Result

2. OR replaces selected bits with one's by masking

1100101	Word
0000010	Mask
1100111	Result

3. EX-OR provides an exclusive OR compare = all zero's

1100101	Word
1100101	Mask
0000000	Result

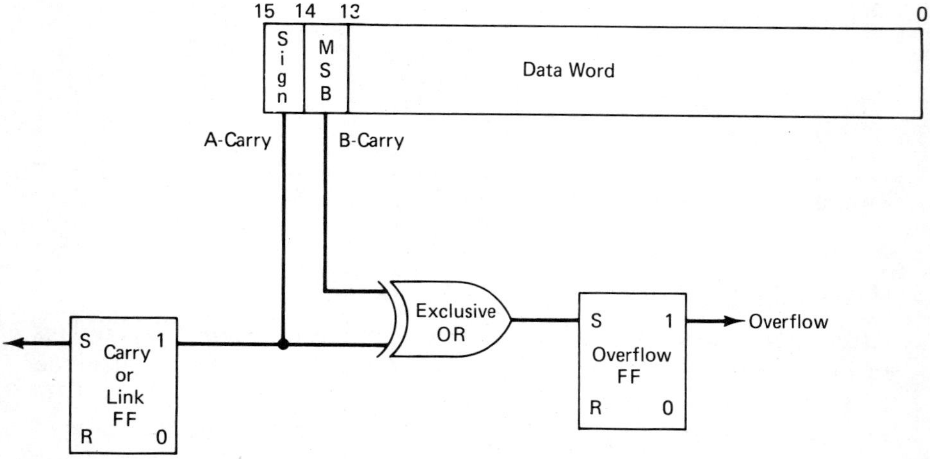

Figure 12-13. Overflow (OVFL) and Carry (CRY) Flags

Shift/Rotate

Shift and rotate instructions are extremely valuable for performing bit checking or for packing and repacking data. The shift instruction alone works well for bit inspection. Suppose we have an accumulator full of bits where the state of each bit has a discrete meaning. The bits could be inspected by shifting them through the carry flip-flop (Figure 12-13) and checking the state of the flip-flop. Shift instructions also provide a simple way to multiply or divide by multiples of two. Each time the data in the accumulator is shifted left, it is multiplied by two, provided bits do not "fall" off the end of the register. Each time it is shifted right, it is divided by two. The rotate instruction provides a way to rearrange or to pack and unpack two bytes (8-bit words) in the accumulator.

The rotate mechanisms are shown in Figure 12-14. If two bytes of data are stored in the accumulator, and we wish to swap them around, either method may be used. With rotate left, the upper byte shifts into the lower part of the accumulator via the carry flip-flop as the lower byte is shifted into the upper byte space.

As previously mentioned, the accumulator is often extended by means of an extension register (ER) as seen in Figure 12-15. Rotation may be used here as well.

Miscellaneous Instructions

1. EXECUTE: The execute instruction tells the computer to execute an instruction located at the operand field address. This does not change the program counter, and the computer executes the next instruction in sequence.

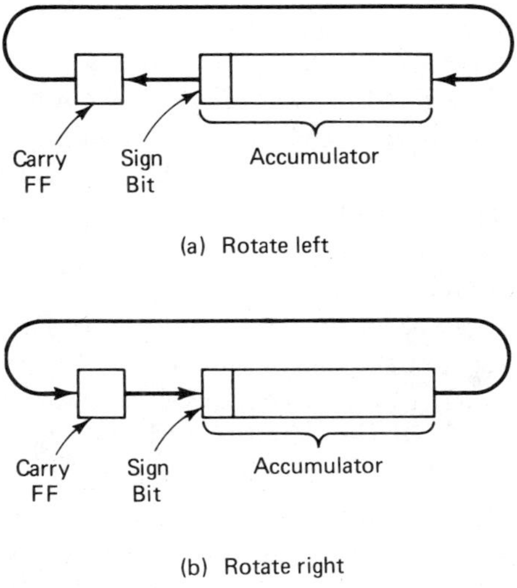

Figure 12-14. Rotate Operations

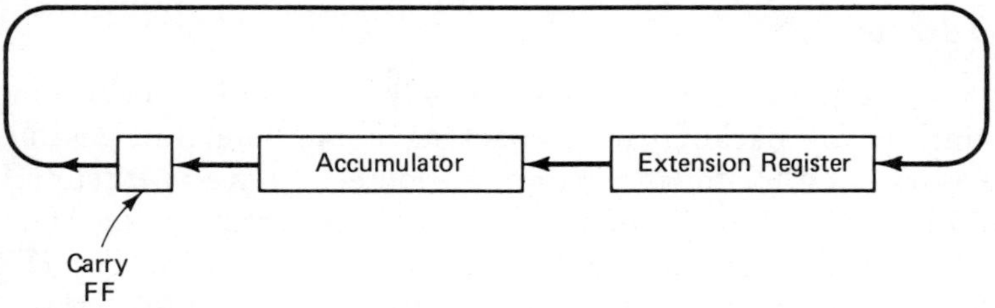

Figure 12-15. The Extended Accumulator with Rotation

2. JUMP: This allows the computer to branch or to move to a new point in the program by changing the contents of the program counter.
3. SKIP: The computer increments the program counter by 2 instead of by the normal 1.
4. NO-OP: Some computers allow a NO-OP or "no operation" instruction. It causes the program to delay.

12-7 OTHER ARCHITECTURES

Although we have alluded to multiple-register computers, most of our discussion this far, for simplicity, has centered around the single-accumulator machine. Most modern minicomputers, however, have multiple registers. Although they can perform all the operations discussed so far, they add another dimension of flexibility.

Such an architecture can be seen in Figure 12-16. As we see, the registers are connected between a set of input gates and a set of output gates. Data is routed in and out of the register file by the CPU control in accordance with the instruction being executed. The memory buffer (MB), memory address (MA), and program counter (PC) are dedicated registers, while R1 through R*n* are considered general purpose. If we had operands located in registers R1 and R3, for instance, we could add them together and store the result in R2, without having to fetch data from memory or destroying the original operands. Any of the three operands (R1, R2, or R3) can be utilized in further calculations and the results stored in other registers from R4 through R*n*. This ability to store intermediate results and reuse them without returning to memory can greatly improve the speed of a computer.

An instruction word for a 32-bit multiple-register computer can be seen in Figure 12-17. This machine has eight general-purpose registers. When register-to-register operations are accomplished, the two registers involved are designated by the A-register and B-register fields. In this case, the results would end up in B-register and the original contents lost. In deference to the instruction formats we have observed

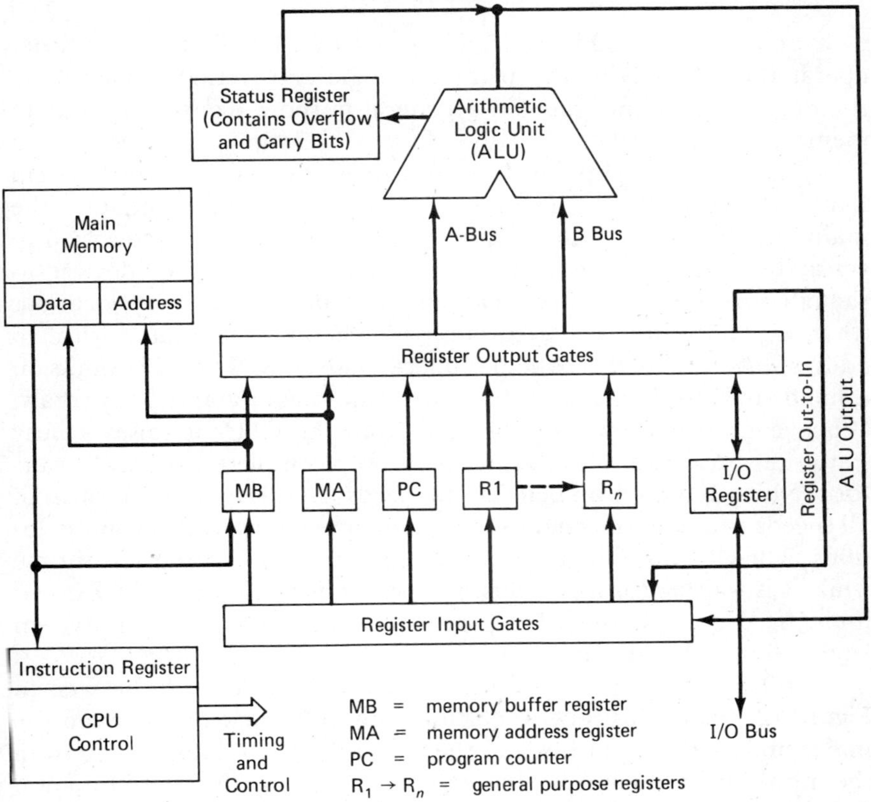

Figure 12–16. Multiple-Register Architecture

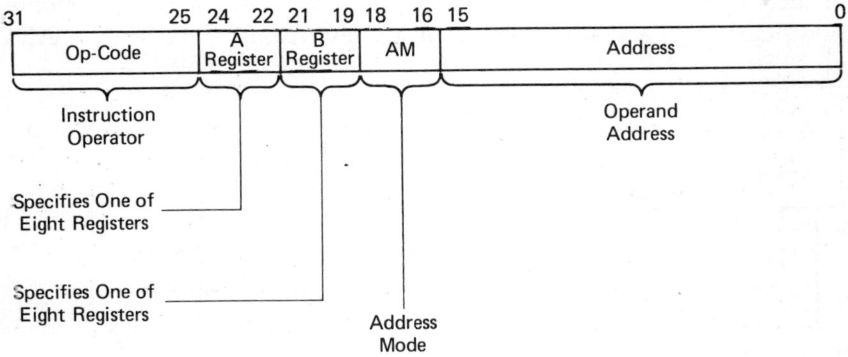

Figure 12–17. Instruction Word for 32-Bit Multiregister Computer with 8 General-Purpose Registers

thus far, the address modes (direct, indirect, paging, indexing, and the like) for this computer are combined in a 3-bit field, which allows up to eight options. This computer allows up to 128 op-codes and direct addressing of 65,536 locations.

Looking back to Figure 12–16, we see that I/O data can be either operated on directly by the arithmetic logic unit (ALU) or passed via the output and input gates to a general-purpose register or to main memory.

Another very popular architecture is the universal bus system found in Figure 12–18. In this design, the central processing unit, the main memory, and all the peripheral devices communicate over a universal bus system. The CPU addresses the various I/O devices as though they were memory locations. The highest memory locations (that is, addresses 64,512 through 65,535 in a computer with 16 address bits) are used for the peripheral addresses. The CPU works on a hand-shaking, interlocked basis with the memory and peripherals. When a device accepts a command from the CPU, it raises a busy status flag and keeps it up until the action is complete. The CPU waits for the busy flag to drop before it proceeds. In this manner, devices of all speeds can be connected to the system without special accommodations. The CPU usually has a "watch-dog" timer to keep track of time while it is waiting for a peripheral to respond. Should the timer maximum be exceeded, the CPU proceeds. This timer keeps the system from being hung up with a faulty device.

Still another type of architecture, the dual bus, can be seen in Figure 12–19. In this system, all functions are connected between the major buses—the input bus and the output bus. All data moving from the input bus to the output bus are routed through a system function or register. In the other direction, data must pass through the modifier. In a fetch operation, the instruction passes through the modifier on the way from the MBR to the instruction register. Any instruction modification is accomplished as it passes through. To do an ADD

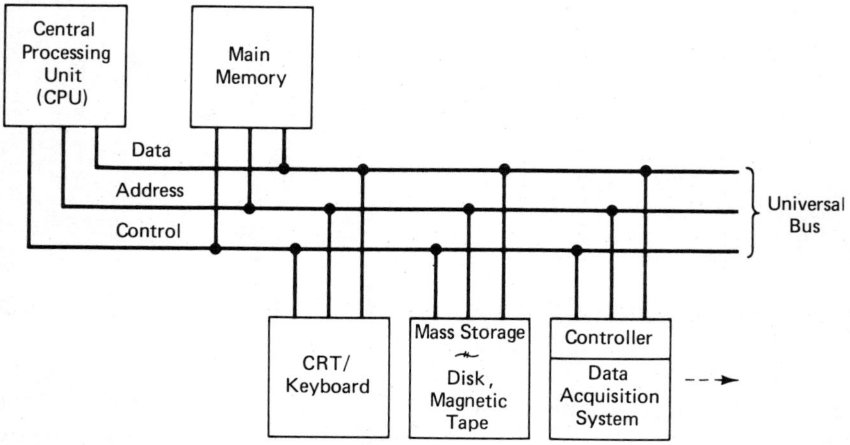

Figure 12–18. Universal Bus Architecture

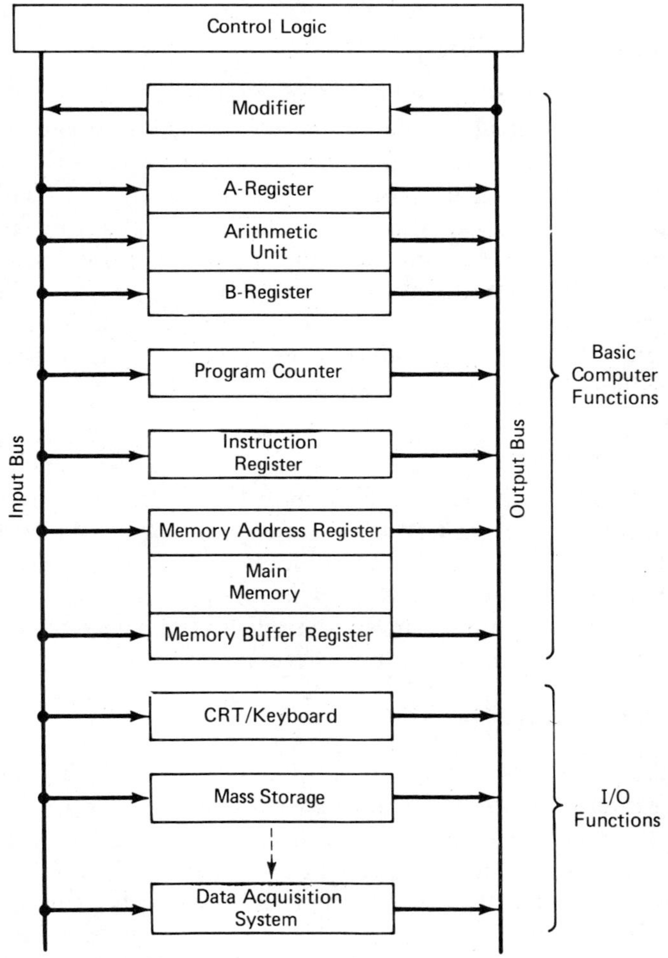

Figure 12-19. Dual Bus Architecture

instruction, the system loads first the A-register and then the B-register from the memory via the modifier. Then the contents of the two registers are added, and the result can be stored either in one of the registers or in memory, but again via the modifier. As all peripherals fit nicely between the two common buses, this architecture works well for controlling external devices.

12-8 INPUT/OUTPUT

Of great importance to the minicomputer system is its input/output system, because it defines the computer's interface to the outside world. Some of the more common structures are defined in the paragraphs that follow.

Multiple Unidirectional Bus I/O

The multiple unidirectional bus I/O can be seen in Figure 12–20. This system utilizes a set of output buses where output denotes data from the computer, and a set of input buses where input denotes data to the computer. The output bus contains data, address, and control lines, while the input bus has data, status, and interrupt information. The status lines inform the computer of the condition of the device (such as busy), while the interrupt informs the computer that the device has completed a task or requires service. This type of I/O generally requires timing information from the computer so that total synchronism is maintained.

Party-Line I/O

The party-line I/O system can be seen in Figure 12–21. Here the input and output data lines, as well as the input and output address lines, are all combined. Some control lines go in one direction while some in the other. For instance, the status and interrupt lines are inputs to the computer, while command lines are inputs to the device. The universal bus falls into this family of I/O structures.

Radial I/O Bus System

The radial I/O system can be seen in Figure 12–22. In this system, each peripheral device has its own special connector to the computer, and its interrupt priority is established by the order of connection. Each device independently talks to the computer. In effect, a device controller has been attached to each port to buffer the computer's internal bus system. Computers built especially for data acquisition often have this type of I/O structure so that each DAS peripheral has its own I/O port.

Direct Memory Access

Most of the I/O systems discussed to this point require central processing unit intervention for each word of data transferred. Another type of computer I/O can load from or store to memory large blocks of data with minimal control by the CPU. This I/O system is referred to as direct memory access (DMA). The DMA usually has its own memory buffer register, as well as registers to store start address and stop address. When starting the DMA to retrieve a block of data, the computer loads the start address register with the location of the first data

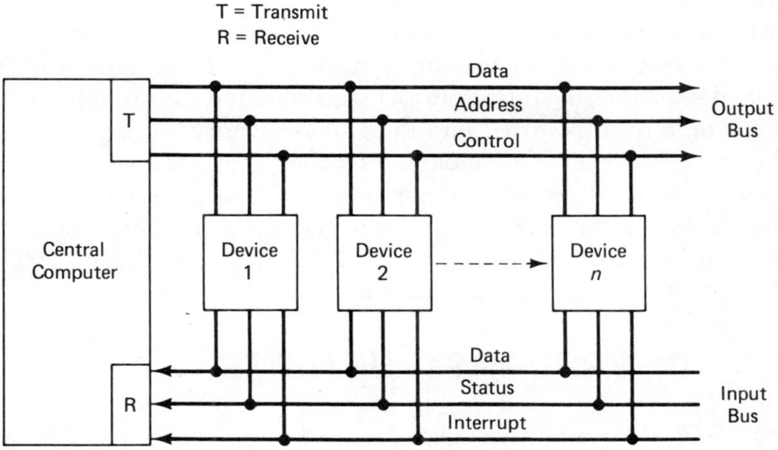

Figure 12-20. Multiple Unidirectional Bus I/O

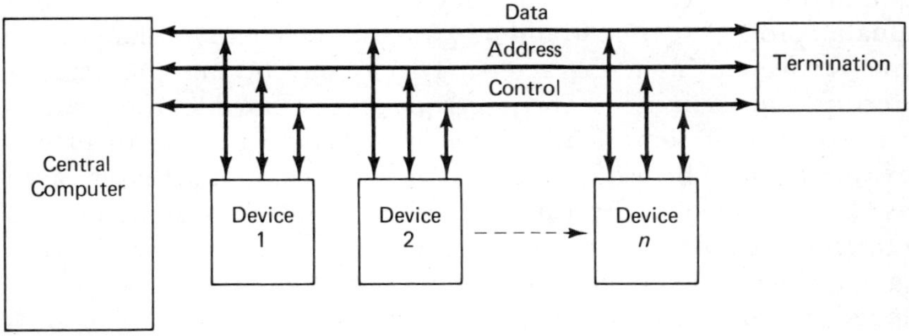

Figure 12-21. Party-Line Structure I/O

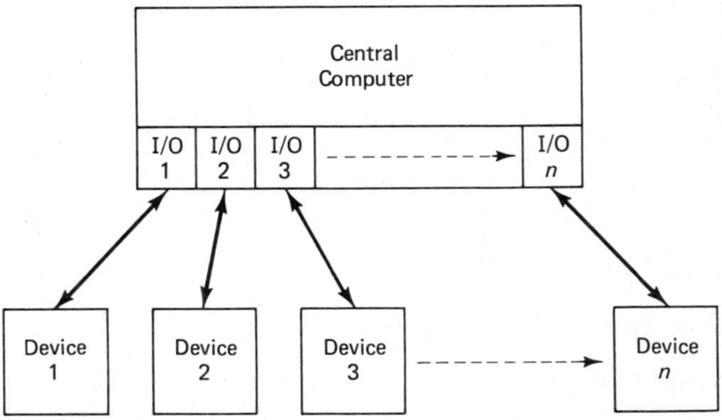

Figure 12-22. Radial I/O Bus System

word and the stop address register with the address of the last data word. When given the command to start, the DMA automatically proceeds to read the memory from the start location to the stop location. Each time a word is stored into the DMA memory buffer register, the start address register is incremented by one. After each word is loaded, the start address register is compared with the stop address register for a match. When the match occurs, the DMA action is terminated. The DMA works independently from the CPU and periodically "steals" a CPU cycle to perform its operations.

12-9 INTERRUPT STRUCTURES

Very often the computer must be interfaced with very low-speed peripheral devices, such as the teleprinters discussed in Chapter 11. The devices often have data rates as low as 10 cps, which means the computer can service the output device only once every 0.1 sec. Much valuable processing time would be wasted if the computer had to mark time between printer characters. To circumvent this problem, the interrupt process is used. The interrupt process provides the computer with notice that some action must be taken to service an external device, or it may be used to simply notify the computer that an action has been completed. By this method, a number of asynchronously occurring external events can alter the computer's instruction processing sequence. The relative importance of the various interrupting inputs determines the peripheral's position in the interrupt structure or priority hierarchy.

Interrupts take time to service. Data that is associated with the program in process must be stored away before the computer can jump to the interrupt service routine. The time required depends on the number of general-purpose and special-purpose registers the computer contains.

To evaluate computer capability, you must understand the type of interrupt structure, the number of priority levels, and the time it takes to service an interrupt request. So let's describe some of the more common interrupt structures:

Conditional Interrupt Instruction

The interrupt is allowed only after the execution of an interrupt enable instruction. At points in the program, the system is checked to see if interrupts are present. This scheme generally has only one interrupt line and all users are "party-lined." If an interrupt occurs, the computer must poll to see who is interrupting. The software overhead is relatively high, since the interrupt enable instruction must appear periodically throughout the program to ensure fast service.

Single-Level Interrupt

The single-level interrupt is similar to conditional interrupt with the exception that the interrupts can occur without an interrupt enable. Users are "party-lined." When an interrupt is detected, the users are polled. The priority of a user is established by the order in which he or she appears in the polling sequence.

Multilevel, Single-Port

With this concept, a test signal is sent out to the users when an interrupt occurs. The users are connected in a daisy-chain fashion. A user who has not requested service passes the test signal message along. Users who have requested service stop the message and place their address on the input bus. This system allows computer control to be transferred to a particular memory location where an interrupt routine assigned to that specific device is stored. The peripherals are attached to the daisy chain in order of interrupt priority assignment.

Multilevel, Multiport

In this interrupt structure, each device has its own specific interrupt line assigned to it. When an interrupt occurs, a flip-flop is set in the computer, and the computer immediately knows which device requires service. Priorities are assigned by ports or physical connection to the computer. This is the fastest form of interrupt structure. Usually interrupts can be masked on or off under software control.

Very often, a second interrupt occurs while the first is being serviced. If the second interrupt is of a lower priority than the first, it must wait. If, however, the second is of the highest priority, the first interrupt is interrupted. Interrupt data is often stored in a last-in/first-out (LIFO) stack of registers. When a priority interrupt occurs, register data from the previous program is stored, word-sequentially, in the stack. After the interrupt has been serviced, the computer restores its registers from the stack and goes back to executing the interrupted program.

When multiple interrupts occur, several blocks of register data may be in the stack, with a pointer (register) indicating the top of the stack. As the computer works its way through the various levels of interrupts, it keeps restoring its registers from the stack top.

12-10 COMPUTER ARITHMETIC

Digital computers generally perform computation with two different types of data formats. These are integer, or fixed-point, and floating-point formats.

An example of *fixed-point words* for a 16-bit computer can be seen in Figure 12-23. The single precision word can express integer words as large as $2^{15} - 1$ or $32{,}767_{10}$. The most significant bit is used to indicate the sign of the number. Very often, 15 bits of precision is not adequate, and the instruction set contains double precision instructions. The double precision data format can be seen in Figure 12-23b. Now integers as large as $2^{31} - 1$, or $2{,}147{,}483{,}648_{10}$ can be expressed.

Another way to increase the value of the largest number the computer can handle is to utilize *floating-point* computation. This implies a floating decimal point and defines the computer's version of scientific notation. The single precision data format for a 32-bit computer can be seen in Figure 12-24a. The highest positive number that can be realized is mantissa $= 2^{23} - 1$ and the exponent $= 2^{6} - 1$. The exponent represents a power of 2, and the radix point appears between the sign bit and the mantissa. If written out, the largest positive number would be:

$$(+0.8{,}388{,}608_{10}) \times 2^{+63}$$

Double precision arithmetic may be accomplished in floating-point format as well as in integer. The double precision data format can be seen in Figure 12-24b. Floating-point arithmetic can be very fast if it is executed in hardware.

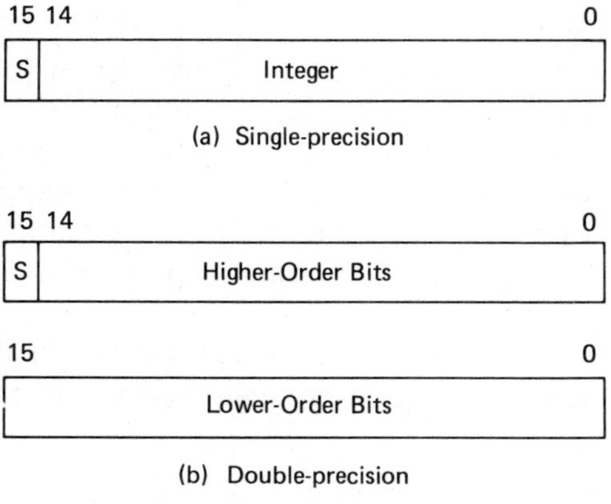

Figure 12-23. Fixed-Point Data Words for a 16-Bit Computer

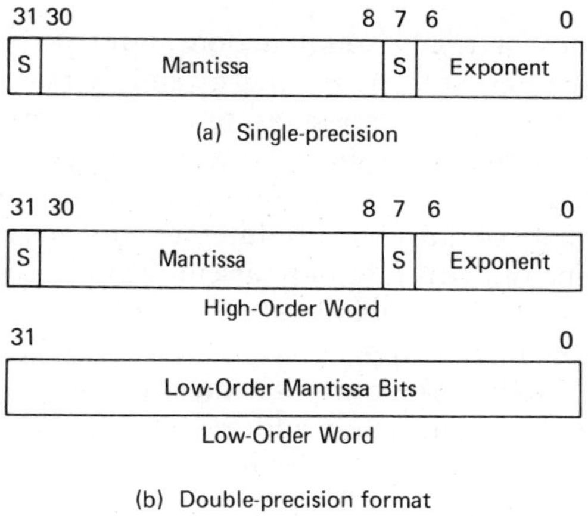

Figure 12-24. Floating-Point Data Words for a 32-Bit Computer

12-11 SOFTWARE

The digital computer has little value if it is not supplied with a program or a sequenced set of instructions to define the operation it is to perform. These programs along with various programming aids are called software. Computers may be programmed using a number of different methods or languages. Let's go through a brief description of language types:

Machine Language

Programming in machine language requires the user to establish the state of each bit in each instruction individually. This type of programming is incredibly tedious and error-prone, even if a 16-bit computer is used and the program requires only a few hundred instructions. With the possible exceptions of debugging a new computer or of making small changes during software debugging, this method of programming is not used.

Assembly Language

Assembly languages are the most rudimentary forms of higher-level computer languages. The instructions are written using simple alphanumeric designations, and each designation bears a one-to-one correspondence to a machine language instruction. The assembler provides all the power of the computer to programmers but relieves them of the clerical task of keeping track of where things are stored in main memory.

The program that converts the assembly language to machine language is called an *assembler*. Many assemblers have what are called macroinstructions. The macroinstruction allows a series of assembly language instructions to be generated by simply making one instruction call. Assemblers can be written to run either on a larger host computer or on the machine of target. An assembler that runs on the target machine is called a *self-assembler*. Assemblers can also be designed to generate code that contains the absolute memory addresses along with the individual instructions (absolute assembler). Or they can generate code that allows address assignment on a dynamic basis. This second type of assembler is said to generate *relocatable code*. The program that loads the code into the computer must be able to "link" the various programs together and assign addresses during the loading process. This program is called a *linking loader.*

Compiler Languages

The compiler language is the highest form of computer language. One compiler instruction may cause the generation of a large number of machine language instructions. The compiler instructions are English-oriented rather than machine-oriented. Some of the most used compiler languages are:

FORTRAN	APL
ALGOL	PASCAL
JOVIAL	COBOL

Often these languages are designed for specific problem areas such as:

FORTRAN	*FOR*mula *TRAN*slation
ALGOL	*ALGO*rithmic *L*anguage
COBOL	*CO*mmon *B*usiness-*O*riented *L*anguage

Theoretically, the language should not be computer-dependent, and, to the programmer, FORTRAN should look the same regardless of the machine the program is to run on. That does not mean that a FORTRAN compiler for one computer will generate code for another. The compiler must match the compiler language to the peculiar machine language of the target machine.

Interpreters

In the case of some languages, such as BASIC, the language converter resides in the target computer at all times. As the computer is fed BASIC instructions, it converts them directly to machine code. This conversion may be accomplished on an interactive basis, and the instructions are executed as they are converted. BASIC is a very popular language in the instrumentation field and for personal computing.

When a computer system is used to perform multiple tasks, it contains an *executive program* or *operating system*. The executive is the system supervisor. It schedules the tasks, regulates I/O, and handles interrupts. It also supervises the loading of programs and allocates memory.

PROBLEMS

12–1. Name the five basic elements that make up a digital computer and describe them.

12–2. The simple computer of Figure 12-4 has the number 150_{10} stored in location 100 of main memory and the number 200_{10} stored in location 101. We wish to subtract the first number from the second and store the answer in location 110. Write down the operations for this simple program in the format used for the examples of Section 12-3.

12–3. A 16-bit computer has the following address modes: (1) direct, (2) indirect, and (3) paging. It can also directly address 1,024 memory locations. How many op-codes does the computer have?

12–4. A computer program requires that a specific group of instructions be utilized several times on different blocks of data located in different sections of the main memory. Which addressing mode would allow the computer to accomplish this most efficiently?

12–5. A 16-bit computer has the number 150_{10} stored in its accumulator and a mask word whose value is 225_{10} in memory. The program calls for the two numbers to be "AND-ed."
a. Show this operation in the form of Figure 12-12.
b. What is the logical result?

12–6. A 16-bit computer has just finished a computation. At this time there are carries from both the sign and most significant bits. What is the condition of the overflow flip-flop?

12–7. Show that shifting the accumulator by 1 bit to the left causes the number in the register to be multiplied by 2.

12–8. Show that a multiple-register computer is more efficient than a single-accumulator machine by means of swapping two words in memory.

12–9. A computer program requires that two blocks of data, each containing 4,096 computer words, be loaded into memory from a mass store. When both are loaded, they must be added together (word 1, block 1 to word 1, block 2, and so on) and the results returned to mass store. Each computer operation such as LOAD, STORE, and ADD takes 1 μsec, and it takes 4 μsec to set up the direct memory access (DMA) unit. If the DMA transfer rate is 10^6 words per second, how long will this operation take?

12–10. A register in a 16-bit computer contains the following floating-point number:

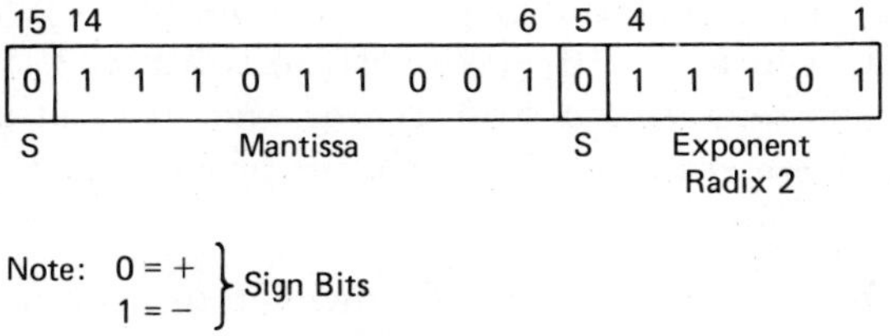

What is the equivalent number to base 10?

REFERENCES

1. G. Korn, *Minicomputers for Engineers and Scientists*. New York: McGraw-Hill Book Company, 1973.
2. J. Lenk, *Handbook of Microprocessors, Microcomputers, and Minicomputers*. Englewood Cliffs, N.J.: Prentice-Hall, Inc., 1979.
3. C. Weitzman, *Minicomputer Systems*. Englewood Cliffs, N.J.: Prentice-Hall, Inc., 1974.
4. I. Flores, *The Logic of Computer Arithmetic*. Englewood Cliffs, N.J.: Prentice-Hall, Inc., 1963.
5. H. Hellerman, *Digital Computer System Principles*. New York: McGraw-Hill Book Company, 1967.
6. C. Ogdin, "Software Design Course," *EDN Magazine* (1977).
7. D. Eadie, *Minicomputers Theory and Operation*. Reston, Va.: Reston Publishing Company, 1979.

Index